The R.A.M.S. Library of Alchemy

Volume 44

Potpourri of Alchemy
Part 1

Compiled by
Hans W. Nintzel

R.A.M.S. Publishing Company

Potpourri of Alchemy
Part 1

Compiled by
Hans W. Nintzel

Produced by

Restorers of Alchemical Manuscripts Society

R.A.M.S. Publishing Company

R.A.M.S. Publishing Company
117 Rutherford Lane
Stuarts Draft VA 24477

R.A.M.S. Library of Alchemy Volume 44:
Potpourri of Alchemy Part 1
Copyright © 2015 R.A.M.S. Publishing Company

First Edition 2015

ISBN-13 978-1519591654
ISBN-10 1519591659

Image Processing by Philip N. Wheeler

This book is sold for informational purposes only. Neither the publisher nor the editor shall be held accountable for the use or misuse of the information in this book.

Printed in the United States of America

Table of Contents

Disclaimer 8

PREFACE 9

Introduction 11

ON THE STONE OF SATURN 13

THE CONFESSION OF TRITHEMIUS 19

ADROP 23

DE URINA 81

WORK ON VITRIOL FOR THE STONE 87

OLEUM ANTIMONII 91

SATURN'S PREPARATION EXPERIMENTED AGAINST LEPROSY 109

THE TERRESTRIAL HEAVEN 117

ASTROLOGY 123

TINCTURA PHILOSOPHICAE 127

NATURES OF SOL AND LUNA 131

PROCESSUS SINGULARIS DE MATERIA CHAOTICA 155

ARCANUM ARCANORUM ARCANISSIMUM 159

A UNIQUE PROCESS FROM PRIMAL MATERIAL 183

THE DIVINE MAGISTERIAL SALT WHICH TINGES ALL METALS
INTO GOLD 187

MIXTURA PRAECIPUA MAGISTRALIS 225

THE MYSTERY OF URINE 235

THE UNIVERSAL PROCESS 245

THE SPAGYRIC ART 253

Tract on the Tincture and Oil of Antimony 279

THEY MADE THE PHILOSOPHERS STONE 301

The Fiery Wine Spirit 347

A Word from the Publisher 353

Dedicated to Hans W. Nintzel,

American Alchemist

and

Founder of the

Restorers of Alchemical Manuscripts Society

(R.A.M.S.)

Disclaimer

Liability: The publisher does not warrant or assume any legal liability or responsibility for the accuracy, completeness, or usefulness of any information, apparatus, product, or process disclosed. The publisher makes no representation as to the accuracy or completeness of the contents of this book and specifically disclaims any implied warranty of merchantability or fitness for a particular purpose. No warranty may be created or extended by written sales materials or sales representatives. You should obtain professional consultation where appropriate. The publisher shall not be liable for any loss of profit or other commercial or personal damages, including but not limited to special, incidental, consequential, or other damages.

PREFACE

In assembling the R.A.M.S. materials, I have come across a good many pieces of information that are not large enough to be a "unit" or "book" in of itself. Nonetheless, the data is of enough value to warrant its saving and publication. These data are often pieces of information vouchsafed to me over the last twenty years, or are bits collected here and there. Or are short tracts from various documents deemed of value. These have been collected and are being presented under the heading "POTPOURRI ALCHEMIA".

In these volumes will be found these minutæ & reproductions of two of the scarce "Golden Manuscripts" produced by my teacher, Frater Albertus. This collection, then, represents personal instruction, data no longer obtainable and those treatises that hold some promise for alchemical experimenters. We hope this melange will be of value to alchemical students and that, perhaps, some insight will be obtained from them.

Hans W. Nintzel

March 1990

Introduction

Philip N. Wheeler

"Potpourri Alchemia, An Alchemical Anthology containing such tracts as: Philosoph Adrop, Oil of Antimony, On Saturn, Urine, etc. etc. Produced by R.A.M.S. 1985," is a compilation of 22 short works on Alchemy. This volume contains valuable information of use to the student of Alchemy.

"Alchemical Symbols," Vol. 21 in the R.A.M.S. Library of Alchemy, is highly recommended to the student who wishes to understand this work.

ON THE STONE OF SATURN

A TREATISE OF CERTAIN PARTICULARS WHEREOF IS TREATED THE PREPARATION OF THE MARKASITE OF LEAD AS WELL FOR THE TRANSMUTATION OF METALS AS FOR THE ALTERATION OF MAN'S BODIES, ETC.

by THEOPHRASTIS PARACELSUS

Translated by J. Hester, 1596, from:

"A Hundred and Fourteene Experiments and Cures"

The Philosophers SATURN is properly the Markasite of Lead, and indeed doth excel <u>Sol</u> & <u>Luna</u>, insomuch that <u>Raymond</u> saith, that in this inferior world, there is no greater secret, than that which consisteth in the Markasite of Lead, insomuch that they which have thoroughly sought out the force and nature of it, have bound themselves together by the Vow never to utter those secrets of nature unto the ends of their lives. For so much as his operation is of such kind, as truly it hath many and sundry Uses, which being duly prepared, doth not only altar and change the filthie and corrupt humours of our bodies, but also can change and transmute by sundry experiments, <u>LUNA into SOL</u>[1].

---

[1] Silver into Gold. -pnw

Take of the Mineral Markasite of Lead XII pounds and grind it into fine powder upon a Marble Stone, as they grind colours with Vinegar; being well ground and tempered, put it into a strong glass, and put thereon a good quantitie of distilled vinegar, and stir it well with a stick, and so let it stand in Balneum Maria[2], then set thereof a blind-head, and there let it stand eight days together, stirring and moving it every day six or seven times; then let it cool; and the vinegar will be of a yellow colour, the which ye shall pour forth into another glass, taking heed that ye stir not the faeces; then put thereon more vinegar, & stir it well with a stick, and set on the blind head and set it in Balneo Maria another VIII days, as ye did before, then pour forth that vinegar being coloured into the other glass. And this order ye shall use so long until you see the vinegar no more coloured: For at the last the faeces will remain in a white mass, like white earth.

DISTILLATION OF THE COLOURED VINEGAR

Then take all that vinegar being coloured, and distill it in Balneo until it will drip no more; and there let thy cucurbit stand three days with the faeces, that it may dry well, then take it forth, and thou shalt find in the bottom of the glass a

---

[2] Water bath, similar to a double boiler. -pnw

white matter, the which take out, for in that whiteness, the redness of the Markasite is hidden, which being prepared, availeth much to make Aquam Philosophorum that they call <u>Ardantem</u>.

THE PREPARATION OF THE FAECES

Take the white faeces or matter, and put it into a distilling vessel with a great recipient very well luted, that the spirits go not forth, and set it in the hot ashes, and give it a gentle fire, and then increase it according to art a day or two, and there will come forth first a white water, and then a red or golden yellow oil, the which is to be kept close[3] with great care.

THE FAECES TO BE TAKEN AGAIN

Then take those faeces and set them in a calcining furnace eight days, then take them forth and grind them finely, and put them in a glass with a good quantity of distilled vinegar and stir them well together; that being done, set it again in <u>Balneo</u> for eight days together, and stir it every day seven or eight times, the more the better, then let it wax cold and settle, and pour away the clear part from the Faeces, but take heed that ye stir not the dregs or bottom. Then pour on fresh vinegar, and set it in

---

[3] "Kept close" means that it should be stored in a tightly sealed container. -pnw

Balneo again, and do as ye did afore, and then pour away the vinegar again, and cast away the faeces; for they are worth nothing.

THE DISTILLATION OF THE VINEGAR

Then take the vinegar which you reserved, and distill it with a gentle fire, and in the bottom thou shalt find a salt, in the which remaineth all the force and strength. Then calcine the said salt again, in a reverberatory, four and twenty hours with a great fire, then take it forth and put it in the glass wherein it was afore, and put thereon fresh vinegar and set it in Balneo. And this thou shalt do so often until the salt leaves no faeces in the bottom; that being done, distill thy vinegar as at the first, and thou shalt find thy Salt prepared liquid and clear as Crystal.

CONJUNCTIO SPIRITUS CORPORIS felicet Olei & Salis

Take the aforesaid salt, and grind it upon a stone, dropping thereon his red Oil by little and little, that being done, put it into a cucurbit luted with *Hermes* seal, and so set it upon a trivet in an <u>Athanor</u> twenty days, and it will be fixed into a red stone, so that ye have the true government of the fire. Then take it forth and grind it upon a stone, and according to the weight put thereunto as many *Letones*, of the <u>Calx</u> of fine Gold, and upon all

16

these put on as much of the white water which ye
distilled afore the red oil as all the whole doth
weigh, and close it up with <u>Hermes</u> seal, and set it
in ashes in a Athanor with a soft fire until it be
fixed, but after it be fixed, there will appear many
colours, the which will turn into perfect Oil and
true <u>Elixir</u>. Rejoice in this, but before ye begin
the work, consider of it and pray.

Finis

THE CONFESSION OF TRITHEMIUS

(Abbott of Spanhein)

From: "Traité des Causes Secondes"

Translated by A. A. Wells

God is an essential and hidden fire, which dwells in all things and chiefly in Man. From this fire is everything engendered. It engenders them and will forever engender them; and what is engendered is the true Divine Light which exists from all eternity. God is a Fire; but no Fire can burn, no Light can manifest itself in Nature without the presence of Air to maintain the combustion; thus the Holy Spirit should act within us as a Divine Fire upon the interior Fire of the heart so that the Light may appear, for the Light must be fed by the fire, and this Light is love, bliss and joy in the eternal Divinity. This Light is *JESUS*, who emanates for all eternity from *JEHOVAH.* Whoever does not possess this Light within him is plunged into a fire without Light; but if this Light is within him, then the *CHRIST* is in him, is incarnate within him, and he will know the Light as it exists in Nature.

All things we behold are interiorly fire and light, in which is hidden the essence of the Spirit. All things are a Trinity of fire, light and air. In other words the Spirit (the Father) is a super-essential light; the Son is the Light manifested; the Holy Spirit is a moving Breath, divine and superessential. This Fire dwells in the heart and sends out its rays all through the body, and thus maintains its life. But no Light arises from the Fire without the presence of the spirit of sanctity.

All things have been made by the power of the Divine Word, which is the Spirit or Divine Breath emanated from the beginning from the Divine fountain. This Breath is the Spirit or Soul of the World, and is called *SPIRITUS MUNDI*. It was at first like air, then condensed into a nebulous substance or fog, and finally transmuted itself into water. This water was at first spirit and life, because it was impregnated and vivified by the Spirit. Darkness filled the abyss, but by, the emission of the Word the Light and the Soul of the World was born. This spiritual Light which we call Nature or Soul of the World is a spiritual body which may be rendered visible and tangible by alchemical processes; but as it is naturally invisible it is called Spirit.

It is a living universal fluid, diffused throughout Nature, and which penetrates everything. It is the most subtle of all substances; the most powerful, by reason of its inherent qualities; it penetrates every body, and determines the forms in which it displays its activity. By its action it frees the forms from all imperfection; it makes the impure pure, the imperfect perfect, the mortal immortal, by its indwelling.

This essence or Spirit emanated from the beginning from the Centre, and incorporated itself with the substance of which the universe is formed. It is the "SALT OF THE EARTH," and without its presence the plant would not grow, nor the field become green, and the more this essence is condensed, concentrated and coagulated in the forms the more stable they become. It is the most subtle of all substances; incorruptible and immoveable in its essence, it fills the infinities of space. The sun and planets are but coagulations of this universal principle; from their beating heart they distribute the abundance of their life, and send it forth into the forms of the inferior world, and into all creatures, acting about their own centre and raising the forms on the way of perfection. The forms in which this living principle establishes itself become perfect and durable, so that they no longer decay nor deteriorate nor change in contact with the air; water can no longer dissolve them, nor fire destroy them, nor the terrestrial elements devour them.

This Spirit is obtained in the same way as it is communicated to the earth by the stars; and this is performed by means of the Water, which serves as vehicle to it. It is not the Philosopher's Stoner but this may be prepared from it by fixing the volatile.

I advise you to pay great attention to the boiling of the Water; do not let your spirit be troubled about things of less importance. Make it boil slowly, then let it putrefy till it has attained the fitting colour, for the Water of Life contains the germ of wisdom. In boiling, the water will transform itself into earth. This earth will change into a pure crystalline fluid which will produce a fine red Fire; and this water and this Fire, reduced to a single Essence, produce the great Panacea composed of sweetness and strength - the Lamb and the Lion united.

Finis

PHOENIX ATROPICUS DE NORTE RELUX - THE REANIMATED
BLESSED PHILOSOPHIC

ADROP

Rescued from the tomb of oblivion, illustrated in vivid colors, in accord with its nature, qualities and its proper preparation.

and

Presented to all inquisitive minds yearning for it, in addition to a little treatise by:

JOHANNES ISAACI HOLLANDUS

de URINA

HOW TO EXTRACT ALL TINCTURES BY ITS SPIRIT.

Translated from the Arabic-Chaldean-French into High German

Frankfort & Leipzig, 1744

PREFACE

ad lectorem

VERY GRACIOUS, DEARLY BELOVED FILI & FRATER
DOCTRINE HERMETICAE & PHILOSOPHIAE SANAE &
IMMORTALIS ALUMNE (Son and Brother and Student
of the Hermetic Doctrine and rational and immortal
Philosophy) (or: Very gracious, Dearly Beloved Son
and Brother of the Hermetic Doctrine, and Student of
the rational and immortal Philosophy).

Here you have, after many Tractatibus Chymicis
(Chymical treatises) de magno Lapidis Benedicti
Philosophiae Ysterio (on the Blessed Philosophy of
the Great Stone), a brief but good and comprehensive
treatise, the like of which has never been seen, on
the Universal high-Tincture work, and the Arcanum of
the philosophic *ADROP* and Materia Chymica, together
with its detailed, perfect preparation which,
because it will preserve you from many labyrinths
and errors that occur in this high work, you should
gladly accept.

But Mercurius Alchymistarum non Mercurius vulgi
nec alicujus Corporis ex Metallis imperfectis, sed
est omnium istorum principium a radix: non est
Corpus Metallicum, sed Spiritus Metallicus

essentialis & Temperatus in qualitatibus suis. (But the Mercury of the Alchymists is not common mercury or anything else out of imperfect metallic bodies, but it is the origin and root of all of them: It is not a metallic body but an essential metallic Spirit, and temperer (or: self-controlled) in its qualities.) And it is only a subtle, clear, pure, and eternal substance of Mercurius, Sulphur, and Sal, and it is a Mercury of Sol and Luna.

For this Philosophic Mercury is the beginning of Sol and Luna, in which Nature begins to produce gold and silver, and it is yet not the Mercury of Sol and Lunar as many thousands have believed so far, but it is a Mercury that is found in the mines and which dissolves gold and silver into Mercury.

This our Mercury is added to gold and silver, because we find in nothing else on earth the metallic power that can cause our Mercury's own Sulphur to be coagulated except in gold and silver; for without those two it cannot be prepared or brought to its end or fixity. Therefore there is in gold and silver the astral influence required to perfect Mercury. But it should be noted that we must add Luna to our volatile Mercury in the first operation. It mollifies (or: softens, melts) the innate Sulphur of Mercury, whereby Mercury is coagulated into gold. For Mercury cannot stand great heat, therefore it must be done with gentle warmth. Because if gold is added to it in the beginning and ☉ requires great heat - ☿ must get as much heat as ☉. Then it changes into a red Sulphur which is not, liquid nor suitable for the Art, because its Sal has been taken from it. And if someone would later try to fix the Rubeus Philosophorum (Philosophic Red) with gold, he would again work in vain, guia non fit transitus ab extremo ad extremum, nisi per medium (because he had not made the transition from one extreme to another with the help of an intermediary).

Furthermore, our ☿ may well be able to give to the metallic quality form and the Perfectionem metallicam (metallic perfection), but it does not have it within itself in actuality but only in potentiality, unless it is ripened by *coction* with the addition of ☉ & ☽. Only then will it have this ability in actuality. Then our metallic fire, will not be colored with gold and silver and made fixed with its fixing strength (force) so can it also not be tinged (colored) or self-permanent (constant, steady) with the water of the imperfect metals mixed: Then our mercury (☿) is very

26

volatile (fugitive, transitory, flighty) and an unremaining behind water, although when it is fixed, take the metallic root of the metals as such, and will be a permanent substance.

Likewise, gold and silver are dissolved in this our ☿. In their Corpore (body) they cannot do it. Even if it (Mercury) is made to flow in strong fire, it does not mix well with the metals and does not color them homogeneously and permanently in their nature. Therefore, one cannot be or do anything without the other. (Note: This paragraph is not clear. I assume that it is about Mercury that the author is speaking and not about gold or silver.)

Now we have said Quid sit noster Mercurius (what our Mercury is), but it is not yet known where it can be found or where it is. For it is indeed in gold and silver, but we cannot open gold and silver in order to obtain ☿ from them, because the opening has to be done with this ☿, then it is *Clavis Philosophorum* (the Philosophers' Key).

Therefore, the philosophers have kept very secret where to look for it, so that hardly one in hundreds will find it. Nevertheless, it is indicated in a secret language. At first, the earth had been created without valleys, mountains, stones, and ores. It was flat and fat, and was only transformed into various colors, ores, metals, and the kind and root of the seven metals by the sun's heat. At first, a smoldering-hot steamy warmth developed which went through the abyss of the whole of the earth, together with the qualities of the four Elements. And because the earth had its innate, watery humidity about it, the color became mixed and changed into a fumus nebulosus (foggy smoke or steam), or a vapor of the four Elements locked in the earth. Finally it increased so much, after each vapor always tried to rise higher and the earth was thrown hither and thither, that mountains and hills were formed. In those mountains that vapory condition is the most moderate and has been mixed by time and best closed in. But in level earth those vapors do not accumulate so strongly and so much. That is why in level earth we do not find as much good ore, because the soil of the mountains is especially slimy, clayer, and fat in their depth. Out of the same vapor that is mixed with subtle, pure earth, the nature of the Mercurius Philosophorum arises. But when now this compound is cleansed of its superfluousness and is boiled, a subtle *ignea sicca substantia* is generated from (a subtle fiery, dry substance) it, and it is Mercurius Philosophorum.

This end, then, dearly beloved reader, sufficiently shows the right way to seek our Mercury, the beginning of our Art. And while this Mercury is found in sufficient quantity where ore is being dug out, it is yet recognized by very few. It is neither gold nor silver, nor the common mercury, nor any of the other metals, nor Sulphur vulgi (common sulphur), nor antimony, nor arsenic, vitriol, marcasite, bismuth, spar, magnesium, cobalt, auripigment, salt peter, or the like, but the philosophers say that it is a vaporous little substance composed of the four Elements, and it is a matter that contains all other metals, which can all be made of it.

While we have shown enough what our Mercury is, and where it can be found, and while it can also be sufficiently proven by the *Lumine Naturae* (Light of Nature) that the *Transplantatio Metallorum* (the transplantation of metals) is possible, I esteem that I have this time said all too much about the main key, which is the *Mercurius Philosophorum*, by which one can attain to the Universal Arcanum L.B. All philosophers have considered this Arcanum the highest and greatest treasure of Nature, higher than which nothing can be found in the world. As to how to proceed further with this high tincture work, you will find enough instruction in the present treatise, and nothing like it has ever been found anywhere else.

This excellent, most useful treatise, which has reached us in a strange way, just as other important select writings that have been kept back for so long, I wished to publish here openly, for the best of you, my reader, and all Sons of the Doctrine, while asking you not to take my work amiss but to receive it graciously and to grant me your favor for the promotion of such arcana and other theophrastic writings (note: probably writings by Bombastus Theophrastus Paracelsus). With this, dearly beloved reader, I commend you faithfully to the fatherly grace and protection of the Most High Spagyrus Trismegistis.

ON THE PHILOSOPHIC ADROP

The intention and goal of all philosophers is to bring about in a short time, outside and on earth, what Nature produces after a long time within the earth, that is, to produce true gold and silver through their Art. But to achieve this, it is absolutely necessary for them to imitate Nature and its effect by the Art. That is to say that they must select pure and clean earth, white and red - which they call their ☉ and ☽ and therewith compose their ☿, and in this they all agree.

Just as long and as much as Nature proceeds or does, till finally the pure earth and ☿ become fixed and in plenty supply, just so must you do if you wish to produce something useful. It is indeed true that ☉ and ☽ are nothing but white and red earth in which Nature congeals the subtle pure quicksilver or ☿, and renders it compact *per minimas partes* (in its smallest parts), and has thus generated two kinds of metal, ☉ and ☽ from it. Consequently, the first thing required is that you should have two kinds of earth, that is, white and red earth, which must be clean, pure, and fixed, and that you should fix in them the two Mercuries, the white in the white earth and the red in the red

earth. The earth and ☿ must be united *per minimas partes* in such a way that they remain thus united in all eternity, that they can pass all tests, and that they can be liquified together to such an extent that they can tinge metals (just as saffron colors

▽) to a white and yellow color, and do this in a rather large amount and in an abundance of tincture, so that you need throw only a little of it upon the molten metal.

For then they attract and bring to the fore the nature that is obstructed and held back, all *Corpora* and *Spiritus* which otherwise, when they are in their own species, are not obstructed and stopped. In this way one can tinge *ad infinitum* and rid the human body of various serious diseases. Whatever power and properties can be obtained in common ☉ and ☽ but not without great trouble and work, and no matter how much labor one may spend on it, they cannot be brought to such great virtue and effect, because the power and vigor, that is, that which gives and distributes life and multiplication to every

species, is extinguished in common ☉ and ☽. If,

32

therefore, you can accomplish on earth what Nature produces within the earth, you may rightly be called a *Philosophus Naturalis* (natural philosopher). But you must understand and take note of the fact that the old philosophers did not build their noblest foundation on common ☉ and ☽, and they have therefore written in their books that this Art costs little, and that a poor man can partake of it and enjoy it as much as a rich man. Which would be wrong, if one would achieve it with common ☉ or ☽, as these are very expensive and hard to obtain by the poor. Certainly, many have used a great deal of ☉ and ☽, because they had not correctly understood this. They also lost labor and effort, not without great harm and disadvantage to their bodies and souls, which is miserable to behold. I have not yet known or seen anyone looking for the philosophers' tincture who did not mix common ☿ (which is the cheater of all alchemists) with common ☉ or ☽, and that is the reason why I do not see any who have achieved or found anything in this Art, but rather do I see those who have become ruined and wrecked by it.

Therefore, be on your guard, I warn and beg you, for although you can make ☉ and ☽ subtle and mix them with the tincture and make some Elixir of them, this common ☉ and ☽ are nevertheless not the right means of the philosophers, because their ☉ and ☽ are two tinctures, namely red and white, which lie hidden in a body not yet perfected by Nature into ☉ and ☽. Therefore you must separate them from their dirty, unclean substance, and unite them and bring them together with earth that is pure and clean, namely red and white, according to their nature.

And these two earths are a ferment of their waters, to such an extent that it is unnecessary to have a ferment for common ☉ and ☽, because all that is nothing but something coming from a body: for *all partes nostri Lapidis sunt Homogeneae & Coessentiales* (all parts of our Stone are homogeneous and coessential) and coagulated, which would not be the case if one were to take and use common ☉ or ☽. Therefore, understand the speech of the great philosopher *Guido de Monte* rightly when he writes and says to a Bishop from Greece whom he had taught this Art: "Take a *Corpus* in which there is pure Mercury, clean and without a blemish, and by nature imperfect. For this *Corpus*, when it is made perfect and is well purified, is a thousand times better than the *Corpus* of common ☉ and ☽." He says further: "In our work there are three kinds of *species*, the Green Lion, the *Asa foetida*, which is a bad-smelling water, and the white steam." He says this in order to deceive the simpletons. For to tell the truth, these three things are nothing but one and the same single thing, *Res una & unius ejesdemque essentia*, to which three different names have been given, according to the three qualities (or: properties) in them.

Consequently, while he calls it the Green Lion, he understands it to be the sun, which causes the world to green *per vim attractivam* (by its attracting power) and rules over the whole world - over everything, even if it is still green because it is still sour and unripe, that is, that which is not yet fixed or perfect by nature, like common ☉.

Accordingly, the philosophers' Green Lion is green gold, living ☉, which is not yet fixed but imperfect by nature. That is also the reason why it has the power to reduce all *Corpora* into their first *Materia*, and to make fixed *Materia* spiritual and

35

volatile. Therefore you may well call it a lion because, just as other animals give way to the lion, all other *Corpora* give way to the might of living gold, which is our Mercury.

The water into which our tincture is infused, is our Luna, and consequently we have two tinctures in our ☿, which can be separated. That it is called *Asa foetida* is so on account of the smell which ☿ has. Indeed, when it has been extracted from its *corpore*, it strongly resembles in smell *Asa foetida*. The philosopher says that the smell is very bad before the preparation of this ▽ and also afterwards, until it circulates in *quintam essentiam* and is well prepared. Then it has a very nice smell and is a medicine for lepra and all epidemics and diseases. Without this living gold, you could never make *Aurum potabile* (potable gold), which is an Elixir for life and for metals.

With this <u>Raymundus Lillius</u> agrees when he says: "We dissolve silver and gold with something that springs from their own root in their species and is coessential with them, yet imperfect by nature.

The above-mentioned Raymundus fixes these two tinctures upon gold-lime and common silver-lime with great trouble and at great expense, which may well be good but is only suitable for great and wealthy gentlemen. But there exists another way, better and easier for the poor, which he calls a white steam. Know then for sure that it is indeed so: For during distillation, a white steam appears before the red tincture, which, when it rises in the alembic, turns the glass white like milk. Therefore they also call it Virgin's Milk.

Wherever you find something written about these three things, understand that it is but one single thing which, as indicated, has three properties.

But I will here discuss a *dubium* (doubt) which confuses the simpletons: Raymundus says that our father shows himself in a filthy, dishonest form, and that he is in all things and in all places. How do you understand this?

Indeed, some are so stupid and of such little intelligence that, hearing that the philosophers write in their books that our father is in all things, they take various substances of which some are bad and base. They calcine them, distill them, and conjoin them, and other like things. This the philosophers severely punish, saying: If you look for the secret of the philosophers in human excrements, you lose your time and find that you have been cheated.

The philosophers also say that he is generated between two mountains, that he is thrown upon the dungheap, also trampled underfoot, that he is generated between man and woman, and that he is in me, in you, and in such like things.

This is the reason why some simpletons, on hearing the like, distill urine, others human excrements, others eggs, others human blood, others old rags, and the like. Finally, one gains as much as another.

But because they are so stupid, one should not wonder that they wish to make ⊙ and ☽ out of things that have never been *ex specie* ⊙ is vel (or) ☽. For no one can give what he does not have, likewise nettles do not produce roses.

How then can we resolve this *dubium*? This you should be sure of: that when the philosophers say that our father is in all places and in all things, they speak the truth. There is no great difficulty

in their words if you consider the matter quite
naturally, since on earth there is neither an animal
nor a thing nor a mineral with a living power or
vigor within itself that could be generated without
natural internal heat and without its *species*. That
is also how the philosophers understand it, namely,
that these species are always germinated by the
internal natural warmth, without which heat you
could not have the least little thing. Therefore, in
this way our father is the *pura Materia*, which is
the nature of gold, and it has a heat that gives
power and vigor and multiplication. By this heat the
father can be taken in his *specie* and can be
multiplied: and that is our secret fire of nature
which our father works in the glass, just as the
natural heat, together with adequate humidity in the
earth, does with the fruit, so that the fruit is
first putrefied and afterwards brings forth in great
quantity and multiplicity.

Therefore, whoever does not know our heat, our
fire, our bath in our glass with a moderate fire
(which is always at the same measure and degree
inside the glass, not outside), our mountain of
dung, our *ventrem equinum* (horse's belly), our moist
fire, etc., will never obtain this Stone nor get
near it. We also have our burnt water, our burnt
wine, our water of life, by which some understand

water of life extracted from wine, oil, or other
Liquoribus. And because that which gives to each
thing its power and vigor is the cause of the
multiplication of each thing in its specie, so you
should also take ☉ or ☽ with which you make ☉
and ☽, which has not yet lost that which gives it
power, strength, and vigor, *vigorem & fortitudinem*,
but that which is alive, warm, and moist, and which
has the might and strength, *potestatem potentiam*, to
reduce all *Corpora ad vegetativan suam naturam*
(bodies into their vegetative nature). Because by
its (that which gives the Vigorem) help, the man who
is dead in his *Specie* and who has no more power and
vigor to multiply by the Grace of God, can again
become alive and multiply (or: reproduce in his
Specie).

I have not yet sufficiently explained how our
father can be generated between man and woman, and
between two mountains. But I will elucidate for you
the secret of <u>Morienus</u>, who says in his <u>Epistle to
Aaron the Philosopher</u> that the *Corpora* taken from
small mountains are the white and clear *Corpus*,
which does not suffer any putrefaction or
disturbance (or: change), and is not subject to
them, and is generated between man and woman. By
these two mountains we understand ☉ and ☽, which

40

are far above us, and which generate silver and gold
on earth through their influence, both of which are

in our ☿ . By man and woman they understand *Activum*

& *Passivum*, that is, *Activa* in our ☿ and, *Passiva*
in our earth.

If now you wish to get the Stone, you can have
it, because it is common to both rich and poor. But
there is a secret in this Art, which leads many into
error, about which there also arises a *dubium*. I
said before that our father is a thing common to
rich and poor alike. Now, however, I am asking, is
there a difference between the father (*id est
Materi*) and the perfect Elixir? To that I reply yes,

because our father is nothing but our ☿ , which is

our ☉ and ☾, our red and white tincture which
each of us can have. The Elixir is different. For

just as our ☿ could be fixed, or could become
fixed, in such a way (namely, on its own earth or on
the earth taken from the little mountains) that it

could become an Elixir, our ☿ could also be fixed
on gold or silver earth, both of which are not
common or easy for the poor to obtain.

So that you may understand the beginning
correctly: our Stone is a common and single thing.
Before the perfect Elixir is made, however, one must
have various things out of which to make it. That is
why Raymundus says that its own earth is not
altogether or always natural. Guido understands this
well when he teaches the Bishop that it is all the
same, that he can take whatever earth he wishes,
provided it is fixed and pure, meaning one should
not worry about the earth, of what substance it is -
with which view Alphidius concurs, saying: "The

Feces from which you have extracted your ∇ , are
not worth anything, are no good, Therefore, you may
well throw them away and mix your Mercury with other
earth that is more subtile."

In order to remove all doubt from the poor and
tell him what kind of earth is best, and which is
most useful to him, *quae ipsi proprior aut
propinquior*, so that his eagle can rise and soar on
its wings: Aristotle calls the earth by its own
name, in common parlance, he says the following:
that it is the end of the egg. By that he
understands the nature of the metal which is Mercury
with its Sulphur, well-proportioned by Nature. Three
things come from the egg, however: the red, the
white, and the shell. We require only one thing,
that is, the shell. This is the end of the egg,

which is the last part made perfect by Nature. It resembles a mountain, and is generated between man and woman. When it is well calcined, it is the whitest and subtlest earth, and the most resistant to fire. It also lasts longer in the fire than all other earths, and it also accepts the tincture, so that you can transmute with it, and thus with the Art, the nature of metal which those who work in this Art do not believe, but only those who have tried and experienced it.

The other earths, which contain a mercurial moisture, do not absorb our ☿ as well as this one, because they contain enough moisture themselves. For the moisture that this earth has, or should have, is multiplied by Nature into white and red, in which there is water and oil - as there is in the blood - (which can be prepared for the medicine but not for the tincture of metals, and burnt with the Elixir of life). This earth is sometimes hated (when its inner matter is putrefied), and then it is thrown upon the dungheap, just as is done with an egg when the pure substance has been eaten out of it. In order to test if this earth would drink or accept my ☿, I once threw a little of ☿ upon this earth, and it soon became fat like fresh cheese. When ☿ had evaporated, the earth turned yellow due to the tincture of ☿ . Therefore, take well care and watch the practice, because people are often cheated in it.

In the name of God, dissolve your *ADROP* in distilled vinegar in B.M., and stir it well with a stick, three times a day. When it has settled, incline the glass and either empty it or pour the liquid off, and again pour fresh vinegar on it. Again stir it, and when it has settled, pour it off. Do this as long as the vinegar is colored, which may take eight days. After this, draw it three times through a felt, until it becomes transparent like a crystal. Evaporate it, then draw the vinegar off per B.M., and do this till the rest is like a mash or rubber (or: gum). Now remove your matter from the vessel and preserve it. You will do this with the said *ADROP* till you have twelve pounds of this gum, and then you will have of this earth the earth and the brother of the earth.

Now put three pounds of the said gum in a *Distillatorium* that contains approximately two quarts. Put the *Alembicum* on it, and seal the joints

well with beer, good eggwhite and flour, well mixed
together on a small piece of cloth. Set it on a
sand-furnace in such a way that there be a two-
finger-thickness of sand around it up to the middle
of the retort. Apply the receiver and give it gentle

△ , so that you can get the phlegma, which is not
worth anything. (Do this) till you see a white steam
rise in the *Alembico*, turning the glass white like
milk. Now change your receiver, which you must well
seal, as this steam or smoke rises violently *cum
impetu*. Increase the fire gradually till you get
oil as red as brood, which is an airy gold *Aurum
Aethereum*, bad-smelling, and Philosophic Gold, the
blood of the Green Lion, our *Ungunentum*, which is a
comfort for human bodies in this life. And in the
same form it is also the *Mercurius Philosophorum*, an
Aqua solutiva, which dissolves gold while retaining
its *Species* - and it has many more names. Continue
the above-mentioned distillation for 24 hours after
the white steam has started to come. Then it will be

perfect, but the △ must finally be increased to
the highest degree. Then remove it, close it well,
so that nothing can escape, then preserve it for
later use.

ANOTHER METHOD

Take six pounds of our *ADROP* that has not yet been dissolved *in aceto* (in vinegar), put it into an earthenware retort containing about four quarts, seal it well, set it in a furnace as if you were going to burn *Aquafort*, put the receiver in front, distill the ▽ or *Phlegma* from it - which is not worth anything - with a gentle △ , till the white steam rises. Now change the receiver, seal it well behind, distill, and increase the fire gradually, the longer the more, and finally increase the fire as much as is required to burn *Aquafort*. Continue for 24 hours, and you will have the Green Lion's Blood, which we call *Aquam Secretam* (Secret Water) and *Acetum Acerrimum* (the most acid vinegar). With it you can reduce all *Corpora* to their first *Materia* and can also purge all human bodies of various serious and incurable diseases.

And this is our △ , which is at all times burning in the same way and to the same degree, outside and inside, and this is our dung, our water of life, our bath, our *Venter Equinus* (horse's belly) which produces wondrous things in the secret work of its *Species*. It tests all Corpora, dissolved and undissolved ones, which the philosophers call warm and moist wine, it contains the fire in *ventre suo* (in its belly), like a fiery water, or else it would have no power to dissolve the *Corpora* into their first matter. This is our Mercury, our ☉ and ☽, which we use in our work.

Now remove the *Feces* - which have become charcoal-black - from the bottom of the retort, calcine them for eight days with a gentle fire, and thereafter again eight days with a stronger fire. Continue doing this till they become white as snow, or calcine them three times in a potter's stone (or: furnace) with greater and stronger fire, till they become white.

When you have brought and reduced the *Feces* to a white calx (or: lime), putrefy and change them to a new whiteness and redness by putrefying them with your ☿, which whiteness and redness they did not have before.

Because the philosopher says: First calcine, then putrefy, dissolve, distill, sublimate, descend, fix it, and wash it often with the water of life; dry it, and copulate, *fac matrimonium* (make a marriage) of the body with the soul. If you can mix these things and bring them together naturally, the ▽ will coagulate when you open the *Corpus*. Then your *Corpus* will die from pain, that is of the bloody flux, and it will change color, as you will see in the clouds after three days. It will rise to the moon and afterwards to the sun, by means of the oceanic sea which is round without an end. When it is in a small town and when it is applied and conjoined, the Art is perfect. This work does not require much expense. Rejoice, however, that you have started it, and be patient and continue with the work to the end.

HOW TO PUTREFY AND ALTER

Put part of the said calx into an ostrich egg and pour enough of your tincture over it to cover it completely. Seal the egg well, so that nothing can escape from it, put it for eight days in a humid and cold place, to putrefy it. After eight days, when the matter is dry, pour again as much tincture over it as before, and, let it stand once more for eight days. Continue thus every eight days until the earth will no longer drink or accept anything. Now let it stand at the same place till it turns black like pitch. Then put it in a natural bed, and let the moisture become fixed with the earth, till the earth becomes white as snow. When it is beautifully white, you can divide it into two parts, keeping one for the white, the other for the red.

Now ferment the first part to the white with the calx, as will be said later, and the red with the ⊙ chalk. If you wish to use this Red for making ⊙, you must reduce it to a red powder, like dragon's blood, just by digesting it with a long-lasting fire. With part of your Mercury you can turn this red powder into an oil by circulation. It will be *Aurum potabile*, *Elixir vitae*, and the metal will be changed into perfect gold.

But I will now teach you a general rule. If you wish to make only a white Elixir, it is necessary that you divide your tincture into two parts. Keep one part for the Red work, but distill the other with a gentle fire, and you will get a white water,

which is our white tincture, our eagle, our ☿, our Virgin's Milk.

When you have these two tinctures, or the white and the red ☿, you must practice with them on their own earth or on the prepared calx of metals. For the philosophers say: One must not worry about the earth, of what substance it is. Therefore, take the said earths, which have been transformed into white and red, as said, and ferment them in the following way:

AD ALBUM (to the White)

℞. Silver-lime and altered earth, *ana*. Grind them well together, moisten and sprinkle it with your ☿, which is called, *Lac Virginis*, Virgin's Milk, till it becomes soft as dough. Now put it into a glass *Sublimatorium*, its *Alembic* on top, and first distill the Virgin's Milk off with a gentle fire, and preserve it well. After this, increase the fire and sublimate everything that can rise around the *Urinal*, just as is ordinarily sublimated. This is our sublimated ☿, made from our transformed and metallic *Corpore*, which is thus made volatile with the help of △, so that it can be sublimated so much that we cannot wonder enough.

Therefore, pound all the sublimate with its *Fecibus* and residue, moisten and sprinkle it again with the Virgin's Milk that you have preserved, distill and sublimate it till everything stays fixed together in such a way that no fire can make it rise.

Thus you will have your sublimated and fixed Mercury, instead of which the unintelligent and ignorant will take the common sublimate that is

sublimated with vitriol and saltpeter, in which they
are very much mistaken.

When you have fixed everything in the form of a
white earth, pour on it an equal amount of Virgin's
Milk, so that it floats on top or rises above the
water, circulate it *in Balneo* till it turns into a
thick oil. After this, dry it to a powder in a stove
of ashes, then calcine it, and after that repeat and
reiterate everything, as was first said concerning
the Virgin's Milk. If you do this often, you can
augment it considerably; and in this way you
can augment it *ad Infinitum* by nourishing and
augmenting it with Virgin's Milk.

Finally, if you wish to make a Projection,
coagulate it to an oily substance or a fat powder
(*in Substantiam oleaginosam, vel pulverem unctuosum
sive unguentosum*). Throw one part upon a hundred

parts of ☿ *crudi*, or another prepared metal, and
it will convert it into good silver, passing all
tests.

Just as I have here said about the White, you

must also do with the Red - with the red ☿ upon
the lime of metals, that is, you must ferment by

sublimation upon transformed ☉ lime, as you did with the White on silver lime.

But take note, you will never get the right silver and ferment unless they have previously been converted from their first qualities by our ☿ , and brought to a new whiteness and redness by means of *Putrefaction*, which whiteness and redness they did not have before.

When they have turned white after *Putrefaction* and are able and inclined to merge with our sublimated ☿ , they unite naturally *per minimas partes* and become fixed together, so that they can never be parted or separated from each other. This would never be the case if only one of these two were fixed and not the other. Then they could not unite *per minimas partes*, because the Spiritus could neither enter the Corpus nor penetrate it.

But when the *Fermenta* have been made spiritual, then the *Spiritus* are conjoined, and the *Corpus*, if it was once perfect and fixed, is eager and has a natural disposition to come again into its first *Fixation*, to accept it again, yes, much more so than a *corpus* that had never been perfect or fixed. When it again accepts the said *Fixation*, it carries with it in its *Specie* all *Spiritus*, which are united with it and are not outside its *Species*, such as living sulphur, arsenic, sal ammoniacus, and other like things.

You may well unite conrmon Mercury with the *Fermento spirituali*, but it will never combine so perfectly that it can stand all tests. Just as our Mercury will not unite with the lime of a ferment that has not been altered.

Therefore, this point of natural philosophy excludes all Whites (*Albedines*) and tinctures that do not come from the right kind of alteration and proceed before the tincture was conjoined and united with the *corpore* and *Spiritu*.

In this connection, Raymundus Lillius has given us this conclusion, saying: "Know, my son, that there exists nothing white or red that Nature has created white or red on earth, that could make the Elixir, unless it has previously passed through the philosopher's wheel (*rotam*)".

ANOTHER METHOD: Variations & Abbreviations with which Raymundus Lullius Experimented.

Take calcined vitriol - which calcines like ash and is *pulvis impalpabilis* (impalpable powder) - put it in a *Urinal*, pour enough Virgin's Milk on it to cover it, close the vessel well with some linen, set it to putrefy in a humid place for eight days, and after eight days give it again just as much of your Virgin's Milk, and continue thus every eight days till it no longer absorbs anything. Let it stand well sealed in this place till you see float on top a crystalline earth, like fresh eggs or roes, which remove from the *Fecibus*. Put the matter into an ostrich egg, well-sealed and glued, in a rather strong ash-fire, so that it becomes fixed. Increase the fire, and continue till it takes on a yellow color. After this, increase the fire again till it - the earth – turns red like dragon's blood. Then, once again pour some of your Red upon it, enough to cover it completely, and coagulate it through Circulation, so that it becomes like oil. After this, dry it to a powder, and throw one part of this powder upon 40 parts of finely molten silver with one part of gold, and all of it will be transformed into fine gold.

If you wish it to get a higher (or: stronger) color, as high as possible, take part of your *Elixir* in the egg, put it in a *Urinal*, pour upon it some of your ☿ , composed of a very strong corrosive (aquafort) made of vitriol-saltpeter. Let the ▽ evaporate with a gentle △ , and the tincture of these two will stay in your *Elixir* and will augment it in *quantitate & colore Eliriris*. If you do it often, it will turn into an oil, and if you dip a red-hot silver-leaf or *Laminam* in it, it will be colored inside and outside. If you melt it with a tenth part of gold, it will turn into gold and stand all tests.

But if you take as much white earth ♂ tis or altered soul as there is white earth of vitriol, and fix them together on altered lime, and thereafter you turn it red and into oil with your composed Mercury (with the Aquafort comps.), you will have the *Great Elixir*, with which you can transmute ♄ , ♃ , ☿ , and all other *corpora* into perfect ☉ . This gold, however, is not suitable as a medicine for the human body. You can accomplish this work in three months.

ANOTHER ABBREVIATION

You can also abbreviate your *Putrefaction* and *conversion* by half, if you make it sharper with your red and white waters. Thus you should fix your sublimated and calcined ☿ , then dissolve it several times in your White and Red, till everything is together converted into ▽ . This water will putrefy and alter the lime of all metals in three weeks, because you have united the two Mercuries, which are two fires, namely, the natural △ , and the △ that is against Nature or unnatural.

HOW TO FIX ☿ UM ♎ ATUM[4]

First, sublimate your ☿, and if there is half
a pound of it, add saltpetert, vitriol, *ana*,[5] (of
each) half a pound. Powder them together in vinegar
till everything turns into a white dough. When it is
thus pounded and white, sublimate it as is
customary. Do this seven times, adding each time
fresh matter, till it all becomes bright and clear
like the sun. Then fix it in the following way:

Put two pounds of it in a phial with a neck
five-fourths of an ell[6] long. Seal it well, set it
in fine clean ashes or sand, so that the round part
be completely covered with the ash. Give it a gentle
fire for a week, increase the fire in the second
week, still more in the third, as much as you can.
Then it will become fixed. After this, dissolve it
in Virgin's Milk and proceed as above. If you wish
to get a little white Elixir in a short time, take
your white composed Mercury, fix it upon silver lime
that has not been altered, and when part of it is
fixed, add some more, which you must do frequently
until the fixed matter melts like butter on a hot

---

[4] Most likely: "How to Fix the Sublimate of Mercury" -pnw
[5] In equal parts -pnw
[6] Approximately the length of a man's arm from the elbow to the
tip of the middle finger, or about 18 inches (457 mm). -pnw

tin. Then you must throw one part upon ten parts of pure ore or copper (*arain*), and you will get good silver for various utensils

You can do the same with your red composed Mercury by means of the said red, sublimated, fixed, and calcined water of Mercury. If you melt it in red water, then circulate it on unaltered gold lime, you will have a good tincture for silver utensils and jewelry.

ANOTHER ABBREVIATION

In a *Circulatorium*, upon one part of lime made of eggs, put some of your red and white water, till the lime is covered. Close it with a glass stopper or a piece of glass that fits and closes well, and seal it well with the following *Luto*:

Take lime, iron, gunpowder filings, powdered filings, crushed glass and honey. Boil everything together and seal with it. Then circulate it in the bath till everything is dried to a powder. Now give it again fresh or new Mercury, as you did before, and continue this till it becomes like another metal, and it will convert that into gold or silver, depending on what the *Elixir* is.

In this way you can alter your Mercury on metallic lime, and you could not find a shorter way. Now put one part of gold lime into your

Circulatorium, adding fixed ☿ , and put on it enough of your red Mercury to cover the lime by two

fingers, then seal the *Botiam* (♉) with a suitable glass by the following *Luto* well-sealed together. Take honey, *Bolum Armenum*, powder of iron slag, well calcined, and boil them together till all is thick and black. But know that the *Lutum* will not be good unless it is well boiled.

Now you must circulate it in the furnace with a gentle fire, and leave it such as till the red Mercury is reduced to a red lime, dry and fixed. After this, add once more some of your Mercury that it is black like oil and will no more drink or absorb anything of your Mercury.

Now take some of this *Elixir* that is like oil and project it on ten parts ☿ *crudi & purgati & puri*. If it has previously been put on coal in a crucible and begins to smell, it will convert all of it into a red, fixed and dry powder. Now put these eleven parts in a *Circulatorium* and do as before, and your *Elixir* will be augmented all the more. In this way you can augment immeasurably, and one part of these eleven parts will reduce and convert a hundred parts ☿ *crudi* into powder, of which each part will convert ten parts of metals into perfect gold that can stand any test.

YET ANOTHER ABBREVIATION

Dissolve your red lime of ☉ and Mercury with a strong corrosive made of vitriol and saltpeter, as is customary. When dissolved, put it in a *Circulatorium*, set it in B.M. uncorked, so that half of the ▽ can evaporate. Then seal the *Circulatorium* as well as possible, fix everything into powder with a gentle △ on ash. Now add once more fresh corrosive, dissolve it as before, evaporate it and fix it into powder on ash. Do this ten times, then your matter will be cooked so much that it will not absorb any more of the corrosive, and it will no longer fix into a powder but stay thick like oil, and it will convert all metals into perfect gold, that can stand any tests, but it is unfit for human bodies, etc.

LAUS DEO (Praise be to the Lord)!

CALCINATIO METALLORUM

ħ & ♄

Put one of these in a big iron spoon with a long handle, set it in a big burning flame-fire (or: reverberating fire) that will reverberate the flame on the metal, then draw the foam - no matter how it is - to the edge of the spoon with another iron like a hook. Do this till everything turns into foam. The flame will cause this foam to become white like flour. Remove it from the △ and let it cool down.

OF ☉ AND ☽ CALCINATIO

Make an amalgam of ☉ or ☽, as the goldsmiths do, so that it becomes like butter. Grind it on a marble - only *per se*, without any *Liquore*, till you no longer see any Mercury. Now put it in a *Sublimatorium*, elevate and sublimate it, take the *Feces*, grind it again with its *Sublimat*, sublimate it again, and do this till the Mercury no longer rises. Then throw all into warm ▽, to draw out the salt which you must distill with a cotton cloth or a wick, as you well know. Subsequently, you must well dry the oxide. It will be *impalpabilis* like flour. Preserve it.

CALCINATIO ♂ & ♀

Sprinkle *Laminas* (filings) ♂ or ♀ with distilled vinegar, and do this till all has turned into rust, letting it dry each time at the air. Now put the foremost metal into a strongly burning △ with a spoon, and when it is red-hot, throw it into distilled vinegar. After this, strain the vinegar into an earthenware dish (or: basin, bowl), evaporate it, and you will get a red earth which you must dry and preserve.

ANOTHER ☉ OXIDE

Another philosopher teaches us how to make ☉ oxide in another way, namely as follows: Make an amalgam with one part of ☉ and 24 parts of ☿. Then your ☉ will be soft like dough and quite subtile. Put it into a phial (*id est Phiol*), in a furnace with white sand, give it strong △ so as to make Mercury boil, and continue thus for five days and nights to make Mercury rise all the time. You must push it down again with a small stick

64

wrapped in a small piece of cloth, and in so doing you will make it go down to the bottom again.

Continue in this way and increase the △ the longer the more, till Mercury is again converted into a red powder like dragon's blood, and is so dry as to make you believe that there is no more Mercury. Then let it cool down, and you will find as good an oxide for the red work as can be obtained.

If you are patient and allow the △ to take its natural course in dissolving it (or to dissolve it, according to whether its nature or its Spiritus requires it), you cannot fail in this Art.

A COMMON AMALGAM

Put four parts of ☿ into a crucible on warm ashes and one part of filed ☉ into another crucible on burning coal till the ☉ turns red. Then throw it into the Mercury in the other crucible. When it begins to smoke, stir it well with a little stick till it is well amalgamated. When this is done, put it into a wooden dish full of water, and you have your amalgam.

CONTINUATIO SUPERIORUM

Until now I have taught you how to make the *Mercurius Philosophorum* red and white, and how you can make it more subtile; likewise, how you must make your oxide, how you must putrefy and convert it

into fresh white earth, which is our ☿ ♎, and how you must shorten the time of the *Putrefaction* and *Alteration*, how you must fix and dissolve, in order to make of them a perfect red and white *Elixir*, and finally, how you must multiply it, namely, when you let them imbibe their own white or red water, and that this is the right way and means, although lengthy, but does not cost much and is the great and common way of those philosophers who have profited by and accomplished something in this Art.

Therefore, because I have disclosed this Art to you, be patient in your work or else it will not be worth it; and above all, fear God, believe in Him, live in accord with His commandments, and help the poor. Then you will find point by point that everything that I have written is the truth, provided you understand it correctly and work well to the praise and honor of our Lord Christ, to whom be eternal honor, thanks and praise.

You must work and proceed in the above-mentioned manner with the red and white *Mercurio simplici* or *composito* upon the jewels of the sea (gold and silver) as you have heretofore done with vitriol. Then it will become the *Great Elixir*.

By the said means of the *Putrefaction* you will be able to convert all things. That is why the *Philosophi* say that the *Elixir* can be made from all things that are fixed on △ , because you mix your red and white Mercuries with these things that are fixed and stand in need of them.

And you can thereby convert into a metallic form certain *Corpora* that have never been metal, such as glass-powder and eggshells, which is the earth between two mountains that is thrown upon the dungheap. For if the eggshell is well and perfectly calcined, it can stand the fire much better than ☉ , and there is in the whole world no earth more subtile, fixed, and better for the poor than it.

This is the intent and purpose of the philosophers, to make the metals here on earth in one day (*percifere Metalla*) which Nature makes in the earth in a thousand years, which seems impossible to many people. The *Philosophi*, however, who melt their *Elixir*, on the glass, say that we must not worry much about what kind of earth it is, and in this way the glass can become malleable (*malleabile & ductile*) and be transformed into metal by means of its transparent fixed tincture.

From this it is easy to perceive and see that it is a *Scientia* that is feasible, and that can better retain the volatile nature of Mercury (*volatile Mercurii potest retineye*) or better, become incorporated with it, than one that does not contain any Mercury and is without any moisture, which is in no metallic *Corpore*, be it as well calcined as it may.

Therefore it is evident that our ☉ and ☽ are nothing but a white and red earth in which the pure Mercury is completely fixed (*per minimas partes*), in all its parts, namely, in such parts as can no longer be divided. In this way we accomplish and bring about by the Art what Nature creates within the earth.

That is why I conclude that the earth can be a ferment in a water if it is fixed in itself, and such a water, if it is pure and clean, can ferment the earth in a white or red tincture without any help from the ferment of common ☉ and ☽. Therefore the *Philosophi* say that their *Lapis* (Stone) is as common to the poor as to the rich, which would not be true if one were to have ☉ and ☽ ferment, which are worth a great deal and are hard for the poor to come by. I am also telling you that the *Lapis* can be found where people least suspect it and that it is so easy to find that anyone could have it if the *Secret* were revealed and known.

But *diversi Philosophi* have *diversa media*. In conclusion of all their summaries, however, I am telling you that our earth imbibes, absorbs and fixes all our Mercury, and our Mercury washes and tinges our earth, so that one is the other's ferment. Because the white Mercury of silver gives a perfect silver tincture, and the red Mercury a golden. Then, when they are fixed together, they make ☉ and ☽ together, without any help of common ☉ and ☽.

Yet, irrespective of all this, we must thank Raymundus who says that he fixes his tinctures in the nature of ☉ and ☽, saying: "We take our tincture from a base (or: insignificant, mean) and bad thing and ferment it with common gold, etc." But his work is costly and dear, because he melts his Elixir on common gold which he alters in two years and extracts two Elixirs from it, as I will teach you, that is, white and red, and he causes the white Sulphur to operate wonderfully without the help of ☽.

If you consider it a good idea, you may begin your work on such a basis, but it will be done at great cost because it is made only for the sake of metals, while it should also be done for the Great Elixir of Life. To make it, our red Mercury must be circulated in *Quinta Essentia* in the following way: namely, with Virgin's Milk, which should be composed and perfectly fixed with an equal amount of

sublimated ☿. Alter the oxide in white Sulphur, bene *fixando & calcinando* (fixing and calcining it well), to destroy the quality of the fire that is against Nature. This water (*id est Aquafont*) is added in order to putrefy and alter it. After this, you must nourish it with your Virgin's Milk, which must be such as and not composed, till it is perfectly fixed. This is to be done with a rather large quantity of Virgin's Milk. When it is well fixed in this manner, you must make it volatile and spiritual once more, and volatile solely by your Virgin's Milk. Then turn part of it into oil or a thick *Liquor* like oil, by *Circulation*, and you will have the perfect Elixir for converting quicksilver and all *Corpora* into perfect ☽.

But make the other part red with your red Mercury, fix and calcine it, then make it volatile once again with your *Fluvio menstruali rubeo*, fix it, and then circulate it into an oil. Then you will have the right ☉ *potabile*, *Elixir Vitae & Metallorum* (potable gold, Elixir of life and of metals).

You can make it much better by fermenting it with *Speciebus*. Then you can prepare your *Great Elixir vitae*. Know that, if you take the red Mercury and add some ☿, which must be sublimated, fixed, and circulated on Tutia, vitriol and iron till it is red, you must circulate all together till it is all converted into oil. Then, when you take red-hot ☽ and cool it therein, it will become lemon-colored, and if you melt it with one part of ☉, you will have good gold to make various utensils from it.

REGULA GENERALIS (general rules)

Whoever wishes to accomplish something useful in this Art, must make his medicine gummous, so that it is easy to melt, that it will melt on a tin like wax and yet not fume any more. Each part is to follow the other in the *Perfection*, and it will dilate within itself, penetrate into the *Poros* of the metals upon which it is thrown, and join them in such a way that they can never again be separated.

But if something remains in the medicine or part thereof that can still be pulverized, it is not truly gummous, and if it were projected on the *Metalla*, it would separate the *poros* and render those metals brittle.

Therefore it is necessary that your medicine be so subtilized after it has been fixed by *Solution*, *Coagulation* and *Fixation* that, if it is turned into an oil, the oil cannot burn, so that your medicine could rather be called a *Species* than a *Genus*, which is only a tincture in a fixed color.

When then you can make your medicine in this way, you can give a beautiful color to all metals that can be worked with the hammer, but not to others.

And know that Raymundus says that the *Corpora* that are dissolved with the *Menstruo naturali* are always the second oxide and not the first. He is in the habit, however, of dissolving this metallic oxide with the composed Mercury, in order to shorten the Putrefaction and Alteration. He calls them the second oxide, the natural sulphur, and *Terram foliatam*, which he then dissolves and reduces to oil with the *Simplici Menstruo*, which is the natural *Menstruum*, as said above, etc.

HOW TO MAKE THE BASILISK
AND OTHER PRECIOUS STONES

Raymundus says in his *Practica verbali* that you should pour our $\triangledown$ *vitae* or burnt wine on metal oxide, so that it swims above it by two finger's width, thereafter set the glass in a bath and leave it in it till all the water is covered with a metallic oxide, which happens because the said water penetrates it and causes it to rise. Now empty the water and oil together into another vessel and preserve it well. After this, pour other fresh water on the oxide and do as you did before. Continue doing this until the oxide is all converted into oil which, as he says, cannot be burnt. Subsequently, set it in a bath, first distill the water off and the oil will stay at the bottom. Now pour the drawn-off water back on the same oil and let both stand together in Balneo for five days. Thereafter, distill everything that could rise with the water and preserve it well. On the earth that remains, however, pour other fresh ▽, and do as before. Continue till all the oil rises with the water, which he also calls *Aquam abundantam & fructuosam*, *Lac Virginis* and *Aquam permanentem* (abundant and fruitful water, Virgin's Milk and Permanent Water). He says further that you must rectify it seven times till it is clear like crystal. In addition, you must *elevate* the *Feces* that remain after every Rectification, and do that with fresh water till everything is *exalted* to Crystalline Water. Now put the ▽ into a phial five-fourths long, seal and stopper it well, and set it in the earth up to the middle, but take care that the stones that may be in the earth do not harm it. Let it stand there for a whole year. Within this time the ▽ will coagulate into a visible clear Stone which Raymundus calls *Basiliscum*. For just as the snake called basilish can kill a man by its mere look, so this Stone is again dissolved into ▽ in Balneo; it will afterwards always coagulate in the cold and dissolve in the heat.

You can do this work with the oxide of all metals, but if you do it with ☽ or ☉ oxide, then - when it is duly dissolved - it is a ferment for all Elixirs you can possibly make, and in one moment it coagulates and fixes quicksilver into fine ☉ or ☽, according to what the oxide had been.

In the same way you can make a multiplication on pearls, and mix these two waters and put them in the furnace, so that it will coagulate into the most beautiful oriental pearls one can find. They are very precious and of great worth, because they are an *Elixir* and precious stones.

If you understand this work rightly, you have the highest work possible on earth. And just as you have made precious stones or pearls from ☽, you can make rubies from ☉ that look much clearer and more beautiful than the oriental ones and which will be the right *Elixir vitae & Metallorum*.

You can also turn the above-mentioned water into a thick *Liquorum* like an oil by fixing it seven times in *Circulatorio*. Then it will become ☉ and ☽ *potabilis*, also a ferment and, *Elixir*.

But to tell the truth, this Art is to be understood quite differently from the way it is described. Therefore, study diligently, and if you do not understand it by what I have previously described, I will teach you but on one condition - because I am just the one who has obscured everything, so as not to communicate it to anyone whom I do not trust explicitly and whom I do not love like myself. For this is the flower of this work, of this Art. And you can work as well with a few ounces of oxide as with hundreds; and when once it has been done, it is not necessary to repeat it.

But you will be able to understand the whole work well by what I have already written, if you can apply one *Solution* with another *Solution*. Because in these words I have revealed to you the whole secret. Understand it if you can, and keep the *secret* to yourself, so that you can give a good account thereof before God on Judgment Day.

The *Corpus* of the volatile *Spiritus*, which is made fixed by the natural fire, becomes volatile again by the action of the natural fire and does not dissolve into cloud-water but into Philosophic Water. Just as the unnatural water dissolves the *Spiritus fixi Corporis* (of the fixed body) into

76

cloud-water, so it congeals *per contrariam operationem* the *Corpus* of the volatile *Spiritus* into crystalline earth, and when the *Spiritus fixi Corporis* is dissolved by the natural fire, it congeals through the aforementioned natural △ into *terram gloriosam* (glorious earth).

The *Corpus fixum* is gold, dissolved by corrosives into clear water, and the *Corrosivum* is unnatural fire. rnstead, if gold is dissolved with the ▽ of ADROP (which is our menstrual flux), it turns back into *terra gloriosa & crystallina* by virtue of our natural flux. When our ☿ is sublimated and fixed with the help of the unnatural fire, which it elevates out of vitriol together with itserf, or draws above itself, then it dissolves again by means of the natural fire, not into cloud-water but into Philosophic Water, which is called Mineral Water. This should be boiled slowly in the Philosophic Egg till it is finally reduced to a perfect red oil, that is, to a red earth, which you must make with our composed ☿ and with *ana* (an equal amount of) Spiritus Vitrioli. That is the mineral Elixir, but not a medicine for human bodies, as above.

FINIS

In the name of the LORD! Make a thing or a substance, but it must first be composed of two that are mixed, so that the smallest part of one is united with the smallest part of the other in such a way that they can never separate from each other. When everything is conjoined and fixed, dissolve again, and continue doing this till finally it turns into a medicine that can be melted and easily driven with the hammer, that is also light and clear, likewise subtile, compact, light and heavy, so that, if set on fire, it will melt like metal before Mercury vanishes and mixes with the said Mercury and with all metals also per minimas - and penetrates them so much as if it were of their own substances and could never be separated from them by the Art after it has become one thing. But it will convert into its own nature all metals or all things with which it has a *convenience*. Because the mixture and *Disposition* is at first somewhat stronger, several and all the more perfect Transmutations will follow. For these arise from the *Mixion* and follow it as the effect follows the *causam*, which, however, can hardly be done with *Sublimation*, because that is a means that causes such a conjunction.

It is necessary, however, that the sublimate boil till it is fixed and becomes fireproof, which will not happen the first time. What is thus fixed must be dissolved once again, and boiled again till the perfection of its melting is sufficient to penetrate the *Metalla* and tinge them. Which, as the *Philosophi* say, normally occurs in the fourth Solution and Coagulation, and it will become really perfect only in the seventh.

Therefore, take natural, that is, sublimate Mercury and a glass or earthenware vessel made and equipped in such a way that it can well stand fire and survive in it. In that boil 3 or 4 ounces till they turn into metal through strong heating of Mercury and completely adhere to the vessel. Now repeat this with fresh Mercury (after the first has attached itself) till the second also adheres. When this has occurred, it must be heated for 5 or 6 days, but somewhat stronger than before. Finally, the vessel containing the medicine must be put into another well covered vessel and given *Ignem reductionis* (reduced fire) till everything that has adhered settles again at the bottom, such as tin or lead, which is the lead, tin, or copper *Philosophorum* (of the philosophers). When this is *generated*, it is then nothing but a woman's work, as the *Philosophi* say. Take the same, throw away its impurity if there is any, and dissolve it in the other part of Mercury, five in six or eight in ten parts. Grind it well for 14 hours till it becomes soft, and put it again into a suitable vessel, let it boil as before, till it is all change into a *Marech*. This must be reduced as you did in the beginning, and when you have extracted it, you must test it to see if it penetrates sufficiently into molten metal or Mercury. Now repeat and *reiterate* your *Solution* and *Coction* till the *Fusion* is sufficient and the *Ingressus* (penetration) perfect, so that it can transmute. And thus, as the *Philosophi* say, you will have their 'egg' and know how to deal with their secrets, etc. etc.

THE END

DE URINA

TRACTUS JOHANNES ISAACI HOLLANDUS

HOW TO EXTRACT ALL TINCTURES BY THEIR SPIRITUM

You must know that all tinctures, white and red, or as they may be, are all made in one way, and it is all a work to the Red and the White. You must fill a large earthenware pot or barrel, well glazed, with old clear urine. Put it on a large alembic together with its receiver, and distill everything you can. Black *Feces* will stay at the bottom. Calcine those for three hours so that they burn softly, then dissolve them in *Aqua communi* (ordinary water), and boil it for two hours. Now remove it from the △ , to let the matter sink. Pour the clear off above and put that back on the △ to boil till a small flux appears on it. Now remove it from the △ and set it in the cold air or in a cold cellar where it will sprout a clear salt.

Remove the salt and again boil the ▽ down, and let it sprout as before. Now take the sprouted salt and dry it in an earthenware pan, for thus it will burn gently without melting. Then dissolve it again in *aqua communa distillata* (ordinary distilled water) and put it back on the △ for a quarter of an hour. Then take it from the fire and let the *Feces* sink; pour the clear off as before. You must always pour the clear off when it is warm, then boil it again till a small flux appears in the form of beans and set it back in the cold air or cellar as before. Receive the sprouted salt and again boil the other ▽ or urine that has not turned into salt till a small flux appears on top as before and it turns into *Sal*. Now dry the salt in an earthenware pan till it is quite dry and preserve it till I tell you how to use it.

Thus you must take all drawn-off urine, distilled by △, and if there is some impurity on the oil or the oil is very greasy, remove it above with a feather or spoon so that it becomes quite pure. Then draw this urine once again off into a glazed earth-barrel (earthenware pot) and *Alembic* with its Recipient (receiver), and do this till no more *Feces* remain in the earth-barrel, and always throw the Feces away, they are of no use.

Thereafter, draw off again by MB, and after that some black *Feces* remain. In this way, draw them off several times till the water goes over clear without *Feces*, and always throw the Feces away.

After this, take the dried aforementioned salt and pour the clarified water into a retort and seal it well above. Set it in hot ashes for three or four days till the *Sal* dissolves into clear water without *Feces*. NB. *Quando Sal solvitur in aquam claram absque fecibus, tunc praeparatum est* (When the salt is dissolved into clear water without Feces, then it is prepared). Then it is done and it is again pure as before, but deprived of its *Fecibus* and its crude *Humors*. And now it has become very subtile, So much so that it cannot be described.

Of this urine you must take six quarters and three quarts of drawn-off *Aceti* and three quarters of *Aqua Vitae*, and half a pound of common salt, half a pound of *Salmiac*, half a pound of common *Calcis vitae* (quicklime). Mix everything well together and let it dissolve into clear water without *Feces*. Then you have a wondrous substance that converts all *Calces Corporum* into their first matter, which is into ☿. With this water one extracts the blessed Quint. Essent. ♄, and from all red and white things. To prepare this water in this way requires ten or twelve operations. Drawn off again, it retains all its power as if it had not been used, but it must be rectified.

HOW TO EXTRACT ALL DESIRED TINCTURES WITH THIS WATER

Take *Sulphur* or *Auripigment*, or ochre or such like from which you wish to extract tinctures. Pulverize them intangibly and then mix them together first with drawn-off *Aceto*, so that they become like soap. Now put this substance in a big receiver and set it on the furnace in ashes or sand. Pour on it clear urine or some of the said water, to half fill the receiver. Seal it above with a cork, stir or shake it well by hand, and incorporate it very well. Then set it again in ashes or sand and first give it a small fire, so that it becomes properly hot. Always remove the cork above and give it air, otherwise the glass will burst. Always stir it by hand as before, so that the matter gets well mixed together and the *Acetum* can well work in it. When you see that the *Acetum* is well colored, pour the clear off above while it is warm, and be careful that no Feces run out when pouring off. Preserve the decanted by itself, well-sealed. Again, pour as much urine upon thee *Feces* as before, and some drawn-off *Aceti*, then seal and shake so that it can mix as before, and when it is colored, pour it off again to the first, preserve it well sealed. You must often pour new urine upon the *Feces* and do as before, till it no longer colors. Then you have drawn all the tincture out of the *Materia*. Now you can throw the *Feces* away or use them as you wish.

Now put the aforementioned tincture into a *Recipient* and remove the moisture till a small flux appears on top. Then let it cool down and pour the matter appears into a closed pot, seal a head on it, and draw off all the moisture in ashes or sand. Then the tincture will remain in the pot red or white, according to the matter, and it is the *Quinta Essentia* of the things from which it has been made. If you have added white, you will find *Quintam Essentiam*. The whiteness must be white as snow, the redness must shine like gold. In the manner prescribed, the *Quinta Essentia Mercurii* ⚍ati can be extracted in the Red or the White; likewise from ♂ or ♀ Nitr. filings, from vermilion (cinnabar) or from *Aere usto*, and also from oxide of ☉ and ☽ or from ♀, briefly, from all things in the world.

Nota: To make the urine stronger, you must again put into it *Salmiac* (ammoniac), and *Sal commune Praeparatum ana 1 dram* (of each one dram), and then extract the colors, as prescribed.

Nota: Of these extracted tinctures you can make *Cementa*, and cement therewith, which is a hidden secret work and Art.

Likewise, you can make of it *Aquafort*, red as blood, shining like a ruby. With this water one can also make wondrous things, which it is not permitted or proper to reveal, etc.

Laus Trismegisto (praise be to Trismegistus)

N. B.

Ein Nössel - one pint

Kalch = Kalk = Lime or chalk, possibly OXIDE

WORK ON VITRIOL FOR THE STONE

A Document from Sir Isaac Newton's

Personal Alchemical Collection.

On Monday, March 2nd., or Tuesday, March 3, 1696, A Londoner acquainted with Mr. Boyle and Dr. Dickenson making me a visit, affirmed that in the work of Jodochus a Rhe, with VITRIOL was not necessary that the VITRIOL should be purified but the oil or spirit might be taken as sold in shops, without so much as rectifying it.

That the fire does not destroy the life of the Oil or Spirit in distilling it from the red hot Vitriol. That two or three pounds of Oil or Spirit will not afford above half an ounce of fixt salt and that, that the Oil affords more fixt salt than the spirit.

That the White spirit is in appearance like rain water, only sweet & fragrant, and that VITRIOLIC Twisdens spirit as I described it to him was genuine. That the white spirit must be rectified seven times from its feces without separating any flegma from it, and that in rectifying it will endure any heat without losing its life. That the remaining matter for extracting the soul must not be calcined to a red heat, but only well dried, lest the soul fly away. That for extracting the soul the spirit must be digested on this matter not two months, but only till it appear well coloured with the extracted soul. That when you draw off the spirit from the soul you must leave the soul not thick like honey or butter but thinner than oil so that you may pour it clean, out of your glass like a liquor and that it will keep better in moisture then when too dry and therefore tis safest to err on that hand then bring no danger in keeping it too moist. I think he said also that the soul must be volatilized by the spirit, but I'm sure 'tis so in the Process of Jodochus, p. 20 & those of Basilius with Gold and other Metals.

He told me also that when all the soul is extracted the remaining matter must be put in a crucible covered with a muffle or hollow cup of iron like a bowl inverted and a fire made round about them for an hour which cannot easily be too hot. Then the salt extracted with the spirit and the matter calcined again and extracted again as before and so on till no more salt can be extracted.

That he imbibed this fixt salt always with 1/8th. part of the spirit (perhaps 8, 10, or 12 times), and that when it was so long imbibed till it became volatile, it was not necessary to sublime it. For all is pure, and if in the sublimation any thing should remain below, that would not be a heterogeneous impurity to be rejected, but an unripe part of the matter which by further imbibition would be all ripened and volatilized like the rest. And that if in imbibing you should at any time use too great heat, all the hurt would be only the loss of so much matter as sublimes and dries upon the upper part of the vessel. And that in every imbibition he let the matter imbibed with 1/8 th. of the spirit continue in the cold for 3 days the better to unite them, and then digested them 4 or 5 days more. And when he had finished the work with the white spirit he imbibed in like manner with eight parts of the red soul (perhaps 7 times). And that when the 3 principles were joined the menstruum becomes a notable one. It then dissolves and volatilizes all metals and gold dissolved and volatilized may be digested with it to the end.

When he had finished the imbibitions, whether with both the white and red spirits, or rather with the white alone, he said that the matter flowed with an easy heat but in cold congealed and grew hard like a stone, and by digestion passed through the colours, black, white, citrine, and red and in the beginning of the decoction and in the decoction it fumed up like a cloud as is described in the process of Jod. Rhe. And that in this decoction if the fire should go out for a while the matter would not thereby lose its life or motion, but go on still when the fire is kindled anew. And that it anew putrefied but in the first decoction, whence I seem to gather that he putrefied with the white spirit alone and multiplied only by imbibing with the red as is described by Jodochus and Basil. The work he fermented by melting with Gold, and said that the whole was finished in 9 months.

THE END.

Casper Oberlein's

OLEUM ANTIMONII

(Oil of Antimony)

Translated by Leon Muller

The Fixation – Its Quintessence

Take 1 lb ☿ii ij, 1 lb sal Tartari (salt of tartar). Grind everything together quite small, put it in a crucible and let it flow (melt) in a wind-furnace for half an hour. Then pour it on a stone and again grind it quite small. Now put it in a glass, pour on it well rectified *Aqua vitae* (aquavit) or Brandy, set it in gentle heat for eight days, and it will turn golden yellow. This then pour off into a clean glass, well closed, and pour fresh Aqua vitae upon it. Then set it in warmth for eight days or till it gets colored again, as before, and it will also turn yellow. Pour it to the other yellow ▽, but pour fresh Aqua vitae on it. Continue with this pouring on and pouring off till no more yellowness comes out. Now distill the yellow Aqua vitae off per Balneum, so that some oil stays behind. The Aqua vitae is always good for drawing out more in this work. When you have drawn off the yellow Aqua vitae down to the Oleum (oil), it will then be the *Quinta Essentia*, as I am quite certain.

After this, take twice as much brickdust as oil. Mix it well, put it into a small alembic (retort), and place it on one side in the furnace, together with a proper receptacle (recipient, receiver). Distill it like an Aqua (water), at first in gentle heat and finally in quite strong heat. The oil will go over quite red. Then it is ready.

After this, take three parts of live ☿ and one part of ☉. Amalgamate them together and pour the oil on it, so that it stands one thumb high above it. Set it in ashes for four or five days, in gentle heat, and the amalgamate will become quite black, coagulated hard, and fixed. Now pour the oil off and keep it. It will always be good as long as it lasts. Now pour fresh Aqua vitae upon the

coagulated mass of ☉ and ☿, and wash the oil off it. It will dissolve into the Aqua vitae. Then distill the Aqua Vitae in Balneo per Alembicum down to the oil. Add this oil to the first oil. It will always be as good as the first. Reduce the coagulated mass with Purras, and what remains is three parts of gold at 24 grades (probably 24 carats).

HOW TO EXTRACT THE QUINTESSENCE FROM ANTIMONY ORE

Take antimony ore, pulverize it quite subtile (subtle). After this, take the very best distilled wine vinegar available, pour it into a cucurbit and add the powder, so that the vinegar stands two fingers' width above it. Stir it well, and let it stand in gentle heat for eight days till the vinegar is colored red. Pour this same red-colored thing off above and be sure to keep it clean. Now pour some vinegar on it as before, and let it stand as before. Likewise, remove its redness together with the vinegar, as before. Continue drawing off and pouring on till no more vinegar is colored red.

Now put the colored matter into an alembic (or retort) and distill it over quite gently per Alembicum. First, the white Phlegma, or water, will go over. Then it will rise above in red drops. Now take the recipient (receiver) away and put another in its place. After this, increase the redness. It is now the right *Quintessence of Antimony*. Use it as you know. It has a sweet taste like honey.

OIL OF ANTIMONY

Item. First of all, the vinegar together with the extracted Antimony must putrefy for 40 days in manure. It must rot in the well-sealed glass. Only after this can the work indicated above be carried out. The oil of antimony is as excellent a medicine for wounds as can be found on earth. Oil of lead is quite like it. You will find that in another little book.

HOW TO EXTRACT THE SULPHUR FROM ANTIMONY

Take 2 Lots of salite, 1 lb of vitriol, 8 Lots of alum. Distill a water from these, as usually, driving the Spiritus hard. Pour the aquafort upon powdered ♁ in a Venetian alembic (retort), so that the liquid stands slightly above the ♁. Let it stand one day, and the antimony will absorb the aquafort. Now pour more of the aquafort over it, so that it stands well above it by two or three fingers' width. After this, distill the aquafort off per alembicum, with very gentle heat and slowly. In this way the ♁ goes over per alembicum together with the aquafort, and the sulphur remains at the bottom white and coarse. It will become so hard that it cannot be taken out but the glass must be broken. If then the sulphur still has some moisture in it, it has to be dried on a lime-kiln. Thereafter the sulphur can be used as is known. Item, the distilled-off water may also be tried for some things. Without doubt, it may be useful for one thing or another.

AN ANTIMONY WORK PARS CUM PARTE

Take tartar, saltpeter, and common salt ana. (The same amount of each.) Grind everything together, put it into a new raw pot, and seal it with a piece of brick. Set it in a coal fire and calcine it well in plenty of heat, approximately a good hour. When it has cooled down, you will find the tartar calcined white. Now powder the whole matter, dissolve it in hot water and discard the feces. Then filter the dissolved matter and boil it down to salt.

After this, take 1 lb of antimony, pound it small, add half a pound of the above-mentioned salt, and mix well. Now melt it in a crucible (Now let it flow in a crucible). Then pour it into a basin, retaining the King. Grind the slag small, add to it half as much salt, mix everything well, melt it as before and pour it to the King as before. Keep the Kings together. Do this work while Kings are being produced.

After this, grind the Kings together. Take fine, small grained ☽, the same weight as the powder of ♂, mix them well and melt them together. Pour this into a basin, and the ☽ will settle into a King. Remove this King, and if you let the antimony smell it well (literal translation), the ☽ will become heavier than before. Now take some more fresh antimony, one pound, and turn it into a King, as before. All these Kings, however, treat with this ☽ as before. And do the same again a third time with a pound of fresh ♄ as before.

When the ☽ has evaporated for the third time by melting on the cupel, take it and put it into another cupel. Add to it three times as much lead as there is ☽, and drive the antimony completely off it. After this, melt it in the cupel till it becomes pure. Now add to this ☽ its weight in fine ☉, and melt them together in a crucible. Then pour the matter into an ingot, laminate it, and cut small plates (or sheets) out of it. Sprinkle them with urine, stratify them with verdigris, not thickly, stratum super stratum (layer upon layer), and seal all together in a crucible with Sal Alcali. Then the silver with the gold has been refined to 24 grades.

Let the slag that has stayed behind melt again and the remaining ☽ will settle into a King. In this way you will get all the Luna back.

HOW TO TURN FINE SILVER INTO GOLD BY MEANS OF ANTIMONY

(That is, one Lot (1/2 oz.) of silver into one mark gold (8 oz.))

First you must cleanse Luna in the following way. Take 3 lbs of rock salt, grind it fine, then mix it with 1 lb of unslaked lime. Now take a crucible, coat it inside with chalk, then put into it one Mark of fine silver, or as much as you wish. According to the amount, and upon one Mark of silver put 8 Lots of the powder ut supra (as above), seal it with a cupel, place it in a wind-furnace, and let the heat start up gently for one hour. After this, heat strongly for three hours to make it melt well till the salt is completely consumed. After these four hours, take it out when it has cooled, and you will find the silver without any water. Do this four times, after which refine the ☽ in lead. Then it is cleansed.

Item. There exist yet much better purgations for ☽ than these. Accordingly, take 8 Lots of cleansed ☿, 1 1/4 Lots of filed ♀, 1 Lot of lead, 1 1/4 drams of iron filings. Melt everything together for one hour. Then take it out. When it has cooled, you will find a beautiful King. Now powder 1

Lot of the King, 1 Lot of sulphur, and 1 Lot of ☉.
Put it in a mortar or throw a burning coal into it,
stir it well, and let it thus burn out. After this,
grind everything. Place this powder in a cement
cupel stratum upon stratum (layer upon layer), the
thickness of the back of a knife, and the above-

mentioned cleansed ☽. Put a cupel over it and
cement it for 8 hours. Then the sheets (thin plates)
will become quite brittle. After this, take all the
matter together as it was in the cement, put it in a
crucible, cover it with Venetian glass and put a
cupel upon it. Let it thus melt in a wind-furnace
for at least one hour. Melt the King on the cupel
for a little while, so that the antimony can
evaporate. Then melt it on the cupel till it is
pure. Separate this, and the lime falls off the gold

to the bottom. Use it as it should. Now the ☽ is at
24 grades.

HOW TO PREPARE THE ANTIMONY FOR

THE ABOVE MENTIONED WORK

Take 1 lb of ♂, put it in a crucible, then pour it into a vessel rubbed over with a bacon skin. In this way its savageness is removed. After this, take the ♂, put it once more into a crucible, add to it two Lots of salite, two Lots of death's head of which Aqua fortis has been burnt (nitric acid), two Lots of red tartar, two Lots of sulphur. Mix everything and put the powder on the melting antimony. Let it melt well for a quarter of an hour, pour everything together into a vessel coated with a bacon skin, as before. Thus the antimony is prepared and can be used for this work. It also serves the goldsmiths to cast ☉ through it, because it does not ravish like the raw ♂. Michael Blaman makes his prepared antimony as follows: He only takes two Lots of lead, two Lots of ♀ and no tartar. With that he refines the gold and uses it (the mixture) to cast through it. I have seen this on the Kuettenberg.

99

HOW TO EXTRACT THE QUINTESSENCE FROM

ANTIMONY OR ☿ ORE

Take antimony, as much as you like, grind it as small as possible, put it in a grass and pour on it very sharp lie made of unslaked lime and ashes of willows. Boil the antimony in it till the lie turns blood-red. Then pour it off into another glass and add another lie, but it must be boiled as before till it turns red. Decant as before. Do this till the lie no longer turns red. Put all the red lie together into a glass, close it well and set it in horse manure for forty days. Then distill with a gentle fire. First, pure water will come out, then drops red as brood. Gather them separately. With them you can sublimate Mercury red, likewise refine and fix. In addition, you can make a tincture with it, as I know myself.

HOW TO MAKE TARTAR RED FOR ANTIMONY

Item. Take two quarters of Acetum acerrimum, distillatum (aceti preparatio p. Mercurii rebedina extrahenda). (This may mean: Vinegar prepared for extracting the redness of Mercury), 1 lb of calcined tartar. Put the vinegar in a grass together with a recipient (receiver). Into that put half a pound of calcined tartar and distill the vinegar over. Now remove the recipient, pour the vinegar into a cucurbit, set it in sand, then distill it per alembicum into another recipient, together with the tartar. Do this alternation seven times. After this, put all the tartar in a cucurbit and distill the vinegar off it. Finally, heat it as strongly as if you were to make Aquafort of it. Now the vinegar is prepared.

OIL OF ANTIMONY

Item. Take 1 lb of vinegar and 1 lb of antimony ground small. Pour the vinegar on the antimony in a glass and let it stand in gentle heat for three days. Thus the vinegar will turn red. Decant it quite carefully, making sure that nothing turbid goes down with it. Now pour some more vinegar upon it, let it stand for seven days till the vinegar turns red. Continue doing this till no more redness goes over. Then distill the vinegar from it per alembicum, and you will find oil of antimony at the bottom.

HOW TO MAKE OIL OF ANTIMONY AND A FIXATION ON IT IN ORDER TO CHANGE 4 LOTS OF GOLD INTO ONE MARK OF SILVER

Item. Take black calcined tartar, 1/2 lb, 8 Lots of ♂. Stir everything well together, then let it melt in a crucible covered with a cupel (test) on it. When it is well melted, pour some sharp vinegar on it, which must be warm, and let it stand till it becomes pure. Then filter the vinegar and you will find the red antimony at the bottom like blook curdled. Turn that into pure dry powder.

After this, take oil of tartar, as much as you like. Dissolve, imbibe the antimony redness with it, and dry it again. Now grind and imbibe it again as before, and let it dry. Do this four times. Thereafter, set it in a humid cellar and it will turn into a red oil called oil of antimony. Let dry what has not dissolved, grind it small and imbibe it again three or four times. Then set it in the cellar and it will dissolve completely.

Item. The other oil is made as follows. Take 4 Lots of vitriol, dissolve it in urine, boil it down to more than half in an unglazed pot. Add to it 4 Lots of ✻, 3 Lots of salite, 4 Lots of ⊕, everything ground small. Boil all of it till it is dry, powder it, set it in the cellar, dissolve it, and you have the other oil.

Item. After this, take the equal weight or mass of these two oils, pound it well with filed ☽ on a hard rubbing-stone (grinding-stone) to make it quite small and subtle. Then put it in a cucurbit and set it to digest for seven days. Pour enough oil on it to cover it a little. When it is distilled over, pour the distilled water back over it. Do this seven times. Thereafter put it in a sealed crucible and let it cement in gentle heat for 16 hours. Then melt it in a wind-furnace for four hours. Finally, refine it in lead, and you will find in 1 Lot of silver a "Quintlin" (4 ounces) of gold at 24 grades.

HOW TO MAKE SILVER FROM ANTIMONY

Recipe (take) 1 lb of antimony, 4 Lots of ♄, 1 Lot of ♀. Melt everything together in a crucible. Pour it into a casting-cone (casting mold). Melt this King a little with a cupel. After this, on the test, and in this way you have the ☽.

HOW TO LIQUATE ANTIMONY FOR GOLD

Recipe. Take 2 Lots of sulphur, 2 Lots of salite. Mix these well together. Now put antimony in a crucible and melt it well. Then throw some of the powder (see above) into it, cover it with a retort (cucurbit). Pour the antimony into a casting-cone, and it will get a King which use as you should.

HOW TO LIQUATE ANTIMONY FOR SILVER

Recipe. Take 1 lb of antimony, 2 lbs of burnt tartar. Put them together into a mortar, then into a crucible. Seal it above and let it melt well. Refine the King as you should.

HOW TO TEST ANTIMONY TO SEE IF IT WOULD BE GOOD TO EXTRACT ITS REDNESS

Recipe. Take the antimony, spread it on yellow paper that has been smoothened with a tooth. If it produces a red streak, it is good. But if it produces a black streak, it is not good. Take, therefore, the ☿ with the red lines, grind it small and mix it with good lime. There must be much more lime than ☿. Pour water on it, let it boil well. Then let it settle and percolate it into a glass. Now it is quite pure. Then add distilled wine vinegar to it, and it will look as if you had poured blood into it. Now let it settle, liquate or draw it off with diligence. Thus you will get a fine red powder which is the Quintessence of antimony.

FINE SILVER FIXED FOR GOLD THROUGH ANTIMONY

Take 8 Lots of ☿, 2 Lots of filed ♂, 2 Lots of filed ♀, 1 Lot of crude tartar, 1 Lot of ♄, 1/2 Lot ♃, 1/2 Lot of glass. These should all melt together. When the matter has melted for half an hour, let it cool so that it may settle into a King. Now remove the King. Grind 1 Lot of the King to 2 Lots of Luna. Melt it for two hours, then refine it in lead. This Luna produces all grades when added to a mixture.

GRADATION

Recipe viridi Aeris, Oleum ✛, Oleum ⊕♄, Pulverm ☿, ana.

Recipe viridi Aeris, Oleum ✛, Oleum ⊕♄, Oleum vel Pulverum ☿, ana. If you can get Oleum Martis, take also one part of that. Put everything together, and when it is coagulated, dissolve it again in a cellar or in a water made for grading (refining). The more often you do this, the more it tinges. If one-third or one-fourth of the powder is thrown into the melting, it will produce ☉.

HOW TO EXTRACT THE REDNESS OF ANTIMONY WHICH IS THEN USED FOR GRADING (REFINING) AND FIXATION

Recipe: Put as much tartar as you wish in a pot, sealed above, set it in a furnace, let it burn till it no longer evaporates, and it will be black enough. Pour warm water on it till it is no longer sharp. Now boil the same lie down to a salt in a kettle or a pot, and preserve it. Take antimony, as much as you like, put it in a crucible, melt it carefully and throw the above-described salt into it, pounded small, so much and so long till the ♂ becomes red. Use it for cementing or grading (refining).

Knock this antimony into pieces and pour some good vinegar upon it. The vinegar will draw the redness out of it, till there is none of it left. It can also be used for grading (refining). But the vinegar that is liquated as described above, is the noblest. If silver that has been laminated fixed is stratified in it, sealed in a glass or pot and closed with a little dough, is kept one day and one night first in gentle then in strong heat. This ♂, thus prepared, is good for cinnabar in which to congeal silver.

(Note: I believe the meaning is: If silver ... is stratified in it, meaning the antimony and the vinegar together and if the antimony is then sealed in a glass etc., then it is good. for cinnabar etc. The German text is not clear. Possibly it means, If silver is stratified in antimony, after its redness has been removed by the vinegar.)

A LITTLE TINCTURE

Take oil of ♄, dissolve in it ✳ made red through ♂. Now take oil of vitriol, dissolve in it ⚹. Take both oils together and remove the fourth part. After this, dissolve sublimated Mercury in the oil, and it will dissolve like water. Now take one part of calcined ☉, imbibe it with three parts of oil (see above). If it is set in warm manure, it will turn into a very red water. Coagulate that in a strong fire and put it back in the manure. When it has again turned into water, coagulate it once more. Do this ten times. In this way it will become a golden water or oil. Now take one part of this same oil or water, throw it upon one hundred parts of crude Mercury, and it will become good gold.

END

SATURN'S PREPARATION EXPERIMENTED AGAINST LEPROSY

OF HUMAN AND METALLIC BODIES, AND FROM WHICH AN OLYMPIC SOLVENT CAN BE MADE

Joseph Duschesne (Quercetanus)

Translated from "Recueil des Plus Curieux Et Rares Secrets" (Paris 1641)

Distil in great quantities a good vinegar as it is the basis and foundation of this work. And, in order to fortify it the better, distil it many times upon its own feces, afterwards mix all that has distilled with equal parts of another not dephlegmated vinegar, and make them pass together so that it may become more efficacious. The fæces which shall remain at the bottom you may put them in a retort at a good fire, by the force of which an excellent oil may be driven, which may burn itself and solve all kinds of minerals.

After having prepared this solvent, you have to take 80 pounds of litharge in powder, and not ceruse, nor minium, nor lead's calx as many artists do especially Isaac Hollandus, take this litharge and put it in several matrasses of great capacity, and pour upon it all your distilled vinegar, so that it swims ten fingers above; then, on an ash fire, you shall extract Saturn's salt by a slow digestion, and upon the faeces which shall remain after the extraction of the salt and of the crystals, you shall pour new menstruum in the same quantity as we have said, and you shall continue this until all your litharge has been reduced to crystals, which, to speak properly, are what philosophers call the chaos or the first metallic matter. Upon this crystalline matter you shall pour again, and for the last time, new distilled vinegar, and you shall make the whole to solve at a slow fire, and you shall filter it, so that a perfectly pure and neat menstruum is made, which, having passed through the vaporous bath, shall leave at the bottom of the alembic a matter melting like wax, which hardens at coolness even as it melts at heat.

You shall then divide this melting matter in many alembics, and shall pour upon it new menstruum, little by little, in order to nourish and moisten it little by little, which you shall do putting at the beginning but only two ounces, then three, then five, seven, augmenting it until the matter does not want to receive anything more, which you shall know when the solvent comes as sharp as it was at the beginning, so that each time you distil your imbibed matter you shall take care to continue until the phlegm comes as sharp as before, because it is thus that the child rejects its nourishing milk after its stomach has been filled up.

This matter, being thus prepared and changed into an excellent and precious gum, you shall digest it at the vaporous bath for 30-40 days, until it remains of a black colour and a stinking smell as liquid pitch, and it is from this liquid pitch that you shall drive out in the same bath an excellent phlegm, which may serve as a proper menstruum to extract from the calcinated earth a precious salt, as we shall later on say.

And, on the other hand, from the repeated distillation which you shall do of the said pitch on the sand bath, eventually giving a good fire above and below, you shall drive out, by the ordinary degrees until a most violent fire, a red and very thick oil which, when united with the preceding distillates, shall compose together a water as strong and violent as the one which is driven out from wine, and even of a higher virtue: which is called by philosophers the water of life of Saturn, whose substance is so pure and subtle that one has to keep it in a most closed vessel so that it may not exhalate. To finish the perfection of this

solvent, this water of life of Saturn is to be put
again on a mild bath, in an alembic of a very long
neck, in which the most pure spirit of this water
shall imperceptibly rise the first (. . .) After
this spirit, a lacteous phlegm shall come over by a
more strong bath, which may serve better than the
one of which we have previously spoken, to wash your
calcined matter, and eventually, by a stronger
degree of fire, after having changed the receiver,
you shall separate also a burning water which at the
beginning shall come white and aqueous then red and
oily, but this one shall be heavy and shall remain
at the bottom of the vessel; you may however make it
to pass over by the strength of the greater fire. As
for the earth that shall remain at the bottom of the
retort as a black powder, you may solve it again by
another distilled vinegar, and change it by this
means into a new lapis of a sticky and gummy
consistency, and finally with the aid of the
digestions and distillations above mentioned, you
may take out from it spirits which are marvellously
active and ardent. Some divide this earth into 2,
and although Isaac himself thus show this division,
I think that it is best and briefier to calcine all
the earth together, and to reverberate by a mild
flame until it does become as yellow as gold, and
when this earth is yellow by the cohobation of the

phlegms you may again extract the salt, according to the rules and ordinary operations of the art.

Having attained to the extraction of this rare and precious salt, you shall take the first spirit which you have extracted little by little by several cohobations, and which you have afterwards kept, and you shall pour it upon this last salt, reiterating this imbibition until an ounce of salt weighs three or four of the spirit, so that it has retained the weight of armoniacal salt of this spirit, and that, to finish with, the volatile overcomes the fixt in proportion: if you do work this operation with exactitude, you shall find at the bottom an excellent earth, which you shall sublimate in an appropriate glass vessel, very clear and well-sealed, where you shall have the pleasure of seeing the sublimation of a philosophical mercury in the form of a happy earth, or rather of a fair gipsy, which you shall keep as a matter of great price.

To crown this work, one part is to be taken of this mercury which you shall unite with four of the spirit of which we have spoken above, or with the same quantity of the first burning water, to make with it a solvent of Sol and Luna, even as philosophers have imagined it, capable of reducing them into a spirit, without destroying their bodies nor losing their species; so that from this solution, truly philosophical, you may do admirable works for the health both of the human bodies and of the metallic ones. This same thing may be done both from choral or from litharge, and in this case you shall make from it, undoubtedly, the fairest and most innocent of all solvents.

"Another solvent for gold by the crystals of Saturn"

- Minium or litharge are solved into vinegar and 3 times recrystallised.

- These crystals are digested 3 or 4 days with rectified aqua vitae.

- The aqua vitae is distilled and a honey or oil remains behind.

- This is the congealed into crystals 6 parts of which are grinded with 1 part of calcined gold.

- This is digested for 40 days.

- It is distilled: first an extraceous humidity comes over, then an oil and many white fumes "which the philosophers call menstrual".

- By rectification, first an aqua vitae comes over, then the phlegm.

- The water of life is poured upon the faeces of the oil which remained behind, and this is left 2 days at the bath to digest; it does become coloured.

- By distillation and cohobation the whole of the tincture is extracted.

- The water of life is retired off at the Balneum "and you shall find at the bottom an oil of gold, most excellent for health when it has been aromatised, with an oil of cinnamon or in other ways."

Finis.

THE TERRESTRIAL HEAVEN

EXCERPTED FROM THE WRITINGS OF EUGENE CANSELIET

Translated from French by: *Gregory S. Hamilton*

There is a sole corporal Spirit, which Nature created first, which is common and hidden and is the precious Balm of Life, which preserves that which is good and pure and destroys that which is impure and corrupt. This Spirit is the end and beginning of all creatures; triple in substance, it is composed of Salt, Sulphur and Mercury, or pure water, which on high coagulates, unites, joins and waters all the lower regions with a fat dry, water.

It is proper and seemly to have received the form and shape which it did, and which Art could not accomplish; by the aid and assistance of Nature it is rendered visible for our eyes.

It hides and conceals in its belly an infinite force and virtue; for it is the one thing which is fully and completely the property of heaven and earth. It is hermaphroditic and nurtures all things, mingling with them indifferently, inasmuch as it contains within itself the seeds of the Ethereal Sphere. Because it is full of a subtle and powerful fire, and descends from Heaven, it has an effect over and imprints its force on the earthly bodies, and its belly, which is very porous, is very hot, and the father of all things. This belly then replenishes itself with more of the vaporous fire, and without cease it receives its nourishment of radical humour which, in the vast body, clothes itself with the water of minerals, which it produces by the digestion of its burning fire.

This Water, which can be coagulated, and which generates all things, becomes a pure earth, which by a strong union holds the virtues of the highest heavens enclosed within; and because in this same earth it is united and conjoined with heaven, I give it this beautiful name: The Terrestrial Heaven.

In the same way that at the beginning the first cause made use of separation to bring order and arrangement to the confused and chaotic mass, Art, which loves perfection, imitates Nature. Nature removes substantial impurity, either by an earthy silt, which it converts to water, or by digestion. Art makes use of purification and digestion, either by water or by Fire, and separates the filth and impurity, purifying and cleansing the spirit, of all blemish. He who knows then the way to use water and fire knows the real path that leads him to the highest secrets of Nature.

The Water, that grand substance, that first creature of God, which is replete with the spirit of fecundity, is the origin of all forms and seeds; and in vivifying by motion it animates all things, and produces all things by the light of Heaven and Earth. This water is the nourishment of all things that live in the two realms: in the Earth, it is a vapor; in Heaven, it is more properly a fire, triple in its substance and first matter; because in a threefold manner and of a threefold nature all bodies proceed from and differentiate themselves from Nature. It contains a Balm which has for its Father the Sun and for its Mother the Moon. Through the air it radiates to the lower plane, and it seeks the high levels and stately strongholds; the Earth is nourished in its fiery belly, and it is the cause of all perfection.

Great God, who gives life to all, has established two medicines for the Spirits and for the Bodies, that is to say, two things which cleanse and purify them of their impurities, and are the causes of disposing of corruption and rebirth to a new life.

The Metals have two things in them, and the two things are the causes of restoration, and they partake of Heaven and Earth, in order that they might unite, and join together the two extremes.

This is why the two are descended from Heaven, and afterwards return to Heaven, in order to manifest their power over the Earth.

As the Sun penetrates the clouds and illuminates the earth, in this manner the Spirit being prepared of this sort and separated from its clouds, illuminates all which is obscure.

Within this spirit there are two forms to consider, in its moisture and in its poison; its moisture is double, and conserves all bodies, with a bitter salt; its poison is likewise double and consumes and destroys them.

Those are the faculties that are shut up in the veins and in the 'cahos', that have the same effects when you take it from the earth; but when it is prepared by the separation of the good from the bad its force must manifest itself and its power over the perfect and the imperfect.

RIDDLE

I live in the mountains and on the plain; I was father before there were sons; I have engendered my mother, and my mother has carried me in her womb and given birth to me without need of nourishment.

I am hermaphroditic and I have two natures; I am victorious over all the strong, and I am unvanquished by the most feeble and small; there is nothing so beautiful under heaven nor anything with a form so perfect.

There proceeds from me an admirable Bird, which from its bones, which are my bones, makes a small nest where, flying without wings, it re-vivifies itself in death. By Art, surpassing even the abilities of Nature, it is at last transformed into a King, which surpasses infinitely the other six.

This is the true Miracle of the Terrestrial Heaven by the Art of the Sages.

ASTROLOGY

Being a tract on the Planetary Influences on the Extraction of the Oil of Metals

By Hans W. Nintzel

Students of the Paracelsus Research Society working on the preparation of oils from various metals often asked: "Why do the results of extractions differ so much from time to time?" Many oils can be extracted in a matter of hours, whereas in other cases it may take weeks or months or longer. In one particular case, there was no visible result after a period of two years! One can then safely assume that no oil will ever appear.

These unsatisfactory results lead to several hypotheses based on topical astrology: The exact moment of time during which the menstruum was poured onto the metal was recorded in a number of cases, and the corresponding horoscopes were erected. Such a horoscope may be compared to a horoscope of conception, a "natal horoscope". For it is certainly not the moment of birth of the oil, but rather the moment at which the whole operation commences, or is "born".

Similarly, our lab records were searched for such experiments where the exact temporal starting points had been recorded, and the corresponding horoscopes were calculated.

We started by erecting the midnight horoscope of the day in question as well as the horoscope of the exact time of conception (i.e., the pouring on of the menstruum). But it soon became clear that in most cases it was sufficient to erect a simple solar horoscope where the first house is formed by the suns sign. This horoscope was easily obtained by consulting an ordinary planetary ephemeris.

We then proceeded by listing all relations between the planetary positions among themselves. But this arrangement was not very conclusive, since it was not possible to draw a unique conclusion about possible relationships between the reacting substances and their horoscopes. The position of the planet Vulcan[7] was then added to our list. Carl Stahl's Ephemeris for Vulcan was consulted. At this point a clear pattern emerged. The best way to understand what follows is by means of analogies. For the correspondences are easier to interpret as analogies, e.g., one can consider that the Sun represents the soul as well as the alchemical sulphur. The theory of analogy and the introductory

[7] Vulcan is a small hypothetical planet that was proposed to exist in an orbit between Mercury and the Sun. Attempting to explain peculiarities of Mercury's orbit, the 19th-century French mathematician Urbain Le Verrier hypothesized that they were the result of another planet, which he named "Vulcan".

astrology lessons as taught at PRS is all that is needed for understanding the following:

The Sun thus represents the alchemical sulphur. The planet Mercury represents the alchemical mercury and the menstruum. The planet ruling the metal being extracted will represent the alchemical salt. The planet Vulcan represents the "inner fire". As alchemy teaches, the regulation of the fire is of extreme importance. Since the menstruum used is the very volatile and etheric substance as taught in the TERTIA class, the regulation of the fire is the only variable that has to be given some special emphasis in this process. All the other planets are relatively unimportant and need not be considered in the horoscope. The only planets that need to be considered are the planets representing the alchemical sulphur, mercury and salt and the inner fire.

TINCTURA PHILOSOPHICAE

Extracted from Theatrum Chemicum, Vol. 1, page 270
(A fragment of "Speculativae Philosophia")

Translated from Latin by: *Patricia Tahil*

Now, so that my philosophical love for the devoted student shall neither help nor hinder, let me cease adding to the philosophical tinctures of our teacher Theophrastus Paracelsus; by their external and internal use cures may be effected, by renewing the blood. They are the simplest and most powerful of all those prepared by alchemical art, and I have described their method of preparation and administration in the *Chirugia manga* of Paracelsus, so that what was previously said about the philosophic remedies may easily be understood.

THE TINCTURE OF SOL, OR: GOLD

When the tincture has been extracted from Sol its white body remains behind, and the tincture is truly pure and separated from its impurity, that is, from its body; this is a necessary separation. The tincture ought to be clarified and exalted in degree, and doubled five times twice two plus four; it does not go any higher. Then you may give it to anyone to renew their blood to its original state, as will be said when speaking of how to administer it.

HOW TO EXTRACT

The body must be torn away from its metallic nature by agitation with salt water, then the remainder must be washed with sweet water, and then the spirit is extracted with spirit of wine. It rises from the tincture, then remains at the bottom.

EXTRACTION WITH SALT WATER

A water is made from the whitest and purest salt possible, made without the decoction and artificial preparation by which salt is usually made, it is liquefied several times, then powdered and mixed with the juice of radish roots, and shaken up with that, and then it is distilled after it has dissolved, and mixed with equal parts of green blood and all is distilled five times, and in that liquor flakes of gold purified with antimony dissolve into powder. It is washed gently with pure distilled water, till the saltiness recedes, for the salt allows itself to be washed away by itself, and does not mix deeply with the substance, but is separated.

HOW TO MAKE SPIRIT OF WINE

Place a measure of the best pure unmixed wine, red or white, in a circulatory vessel, well luted and closed, and dip the vase in the water bath to the depth of the wine, and allow it to boil for ten days. When you have taken it off the heat let it distill into a cold phial or cucurbit as long as the spirit ascends; it gives its own sign and soon ceases; what follows is eau-de-vie and not spirit.

Pour this spirit of wine on to the residue, which is like a very fine powder, to the depth of six fingers, and having sealed the vessel thoroughly place it in a warm bath, where it is to remain for a month. Then the tincture of Sol ascends into the spirit, and a white powder remains in the bottom of the vessel. These two must be separated in turn, Melt the dust and a white metal may be made from it. Let the spirit evaporate, according to the art, and a sort of dry liquid will remain in the bottom, which you will graduate five times in a retort large enough for the amount of liquid. This gradation is by elevation only, to make the material spiritual, but the quintessence does not allow itself to be spiritualized any more (that is, 2.4.0.), otherwise it burns up if you proceed further.

END

NATURES OF SOL AND LUNA

By Michael Scot
Extracted from Theatrum Chemicum, Vol. V, page 713
Translated from Latin by: *Patricia Tahil*

So that the end and the beginning may agree in every way, let us enquire whether true gold can be made by the Art, or not; and at first it seems that it cannot, because gold being a perfect substance, requires its own proper place to be born in, namely the womb or veins of the earth, just as wine is made in the womb of the vine; therefore things can be made only in their own places.

Likewise, a physical form can only be brought into being where its own particular active and passive parts can be artificially introduced. However, the physical form of Sol is not produced by the heat of the sun, or of such fire as craftsmen use. But it is produced by the heat of Sol, therefore, etc.

In this matter the question to be answered is whether one can create artificially in any mineral, be it Sol or Luna, a reproductive ability that can at once harden Mercury to the hardness of Sol. Let us first declare that, supposing one found such an ability to reproduce, then it could properly be extracted from Sol, and extracting according to (the principles of sound) reasoning.

The second question to be answered is how Sol contains such a power of growth in itself.

The third question to be answered is how it is produced in the earth, and what sort (of earth) nourishes it, and what kind of seed must be sown to make it grow.

The fourth concerns the signs (identifying) the mineral power that these external events introduce into quicksilver.

Finally, all arguments are dealt with.

THE FIRST PART

Concerning the first (question), you should know that such a reproductive ability can properly be extracted from gold. This is proved by St. Augustine, writing on Genesis, where he says: All the elements of the universe are present in physical substances, also certain hidden powers of fertility, which, at the right juncture of time and cause, burst forth into their destined states and shapes with their proper scope and limits; and just as one does not say that the Angels who called these forth were the creators of whatever crops in the earth, although they provide them with the right conditions and opportunities for sprouting.

God is in truth the one creator who plants in each substance its own condition and its method of procreation. This is in the book, ***"The City of God"***, copied in chapter thirty one, in question five.

It is no wonder therefore, that philosophers have discussed those arguments in many different ways. Some call those powers that are attendant on fertility, heavenly. Everything has it pattern in the heavens, etc. Others call these powers elementary, for they suddenly burst into action, like fire, or some other element.

Others call them peculiar mineral powers, for they come into existence out of their own particular mineral. And others call these same powers the roots of Sol, for just as grain is fed by means of its roots, so gold is fed by means of these powers, as will be shown below.

We, however, call them fermented spirits, for they loosen up the substance, that is, the purified spirits, and ferment it; their most natural ability by far is to make other (things) like themselves, and if they are perfect they will ferment and tinge; you require nothing else of them. Should anyone object that the philosophers spoke very

figuratively, we (shall) prove that their
fundamental purpose was this (type of)
investigation.

For, following what all philosophers have
plainly said, we maintain that the aforesaid mineral
power is of a universal nature - I mean that it is
not Mercury intrinsically, nor in its whole
substance, but part of it is (Mercury).

He who sublimes, strives to remove the
phlegmatic wetness and stinking substance from
Mercury, and so it is killed, for if it were not
killed like this, it could not complete the task,
nor be fixed; it is for this that he who
precipitates strives with such a strong heat. It
also ought to be pourable, which is what the man who
dissolves works towards, for if it cannot be poured,
it does not enter, nor perfect; these (powers) are
also removed by dissolving.

Gold is altogether mineral, as appears from its
weight and the way it imbibes Mercury. This was
therefore the philosophers' whole and entire purpose
and desire, for it acquires these remarkable powers
by means of heat and the inspiration of
intelligences, with which error is impossible,
except by chance. Nevertheless, all these things can
appear in Mercury with the help of fire, mental
ability and much hard work.

THE SECOND PART

IDEAS ABOUT THE PRIMAL MATERIAL OF SOL & LUNA

As for the second question, you should realize that philosophers seeing the pure, fixed, fluxible, Mercurial material that is gold, and finding that it entered, but did not perfect, fell into a daze, and many of them denied the Art. Some who were more careful considered the principles of philosophy, saw that it (gold) was neither born nor nourished, for the power of reproduction is excess of food, and tried to insert some sort of nourishing power into gold, so that in this way it could flourish and grow.

This, however, is merely enlivening those self-same reproductive abilities into a fundamental power. For a grain of corn is broken up by the power of Sol, this core being neither wheat, nor barley, nor stone, but more ready to receive the shape of wheat than that of stone. This readiness, which naturally tends towards forming wheat, ought to be converted into the fundamental power of Sol, just as they say.

Gold must be broken up in such a way as to reduce it to its primal material, so that it can then germinate; however, they said that this primal material was a certain Mercury and sulphur, for all metals are created from Mercury and sulphur.

This is false, for sulphur and Mercury are distinct minerals, and gold is never discovered where they are found.

Others said that the primal material of all metals is Mercury, because sulphureous particles are mixed in with it, according to Geber, and this does not seem to be true, for there is never any gold where the mineral Mercury is.

And others rightly said that the primal material of Sol is a moist greasy vapor that partakes of both natures and is found in rocks where the purest gold is found, and that it could not be generated there except by a coarse rising vapor, as the fourth chapter of (Aristotle's) "Meteorics" has it. Now such a vapor as we have mentioned must be raised as one and at the one time, but from the coarse part stone is made, from the greasy part, metallic substances. This theory is true.

The reason why gold should be reduced to a vaporous substance is because we see that all metals are generated from Quicksilver, by means of which they are born. As they say: - A man is born of his father by means of sperm, and generates his son by means of sperm. We likewise see that wheat generates a grain like itself by means of the ear, and gold generates gold by means of a vaporous substance, for it is born as was said. Therefore the philosophers declared that gold ought to be broken up and converted into its primal material, sulphur and Mercury.

REDUCTION OF GOLD TO ITS PRIMAL MATERIAL

One should, however, use an established method, and certain determined ways and intermediates when reducing gold to its primal material, otherwise the substance is destroyed, and not improved. He who is considered a better philosopher than all the others is he who is closest to reason, and has understood the words of the philosophers concerning gold - although it can be worn away, and is earthy, and therefore belongs to the genus of cold, dry things, it may, because of its general composition, be called warm and moist with respect to other metals. It becomes a sort of burgeoning earth, so to say, as nails and hair become a definite part of human substance, for with respect to the whole terrestrial Globe, gold is part of the earth, just as hair is part of the human body.

Consider this, moreover; gold is cold and dry, because it is of the earth, therefore, according to what has been said, it should be converted into a vapor that retains its mineral properties, for vapor is moisture, both intrinsically and by pressure of external events, and therefore it should be altered in nature from its own characteristic nature into another material- with its own characteristic nature, so that if it had been hot and moist, it would be artificially made into something that was cold and dry by nature.

It should moreover, be calcined carefully in the reverberatory furnace, and caused to imbibe sharp fiery waters; this operation, when prolonged, breaks up the moist surface of the gold, and produces dryness in its place; therefore it is hot and dry. But because earthy dryness cannot coexist with the heat of fire the first dryness is destroyed and another, black, greasier one is produced.

Therefore it is dried and calcinated gold, of whose workings Hermes has spoken elegantly, for the stone ascends from earth to heaven, that is, into fire, for that particular earth is fire par excellence. When explaining this in his book "*De mineralibus*", Albertus says that the stone ascends from the earth, that is, it is raised from the earth into heaven, that is, into fire, because gold acquires the qualities of fire from calcination, or roasting, or refining.

This operation is soon finished for the affinity hastens the transfer of properties, as we postulated, for earth has an affinity with fire and substitutes for it at once, because it descends from fire into earth once again.

This does not occur in one leap, for the dry calcined substance is broken up by means of our airy water, or the proper action of moisture, by frequent and repeated irrigation with its own water, so that the dry is broken up and an airy moisture produced. But because the fiery heat cannot coexist with such an airy moisture, it is destroyed and becomes another kind of tempered heat. There is therefore nothing better than heat and moisture, for gold hidden in a mineral is found by means of heat and moisture and easily extracted thence as you know.

That heat must therefore be lessened, because part of it is airy and very sharp, and part watery, and therefore cold. Therefore the heat is lessened and made semi-cold, for which reason vapour is said to be the median between air and water, as the philosopher clearly says in "*Liber Physicorum*".

So the gold returns to its essence, that is, into a vaporous substance, and this is called the prime material of Sol, therefore Geber in Ch. 78 of "*Medicina*", speaking to the craftsman of the third order, says: You have treated or extracted the precious earth in such a way as to make it what Hermes claimed, when he said, Again it descends from the heaven, i.e., from the fire, to the earth, that is, the prime material (descends). He at once confirms this theory when he says that it acquires the qualities of the elements above and the elements below.

Then he declares that the four elements should be extracted; doing this is merely exciting the powers of reproduction or active and passive qualities; many have fallen into endless errors when explaining this.

Now the earthy substance which we have prepared, the earth of gold, is called by the names of all sorts of earths and it is called thirsty earth because it seeks food and drink in a manner of speaking, as we shall mention below, and it is called blessed earth because it gives all good things, and leafy earth, because it is fed like the leaves, and grows, and it has an infinite number of earthy names.

All the philosophers saw that the essence of gold was born of calcination, and solution in water, and final hardening in the heat of the fire. Therefore we ought to complete the task by doing all the things we spoke of above.

WHY IT IS CALLED A STONE

Likewise they called this substance by the name of all sorts of salts, therefore some who did not understand properly, fell into various errors. It is rightly called a stone because it redeems the transitory ways of man. And the operation of converting the solid substance to a spirit is called "stone" by many wise philosophers.

We, however, call that inherent spirituous mineral quality by another name; and so it is clear how the means of reproduction that have been placed in gold can intensify and strengthen, which is the same as inserting the power of growing and germinating in gold, which was the second question. But because all spirits must be kept as "moist bubbles", philosophers have tried to convert this spirituous power into a moist, fatty one by frequent and repeated solution, ds said before, and by calcination.

For the earth itself gives birth by continually drinking in the showers descending from heaven, as Isiah has it, and when it has been dried by heat, it becomes fatty. So all salts and crystals and fusible minerals are made in the same way; and therefore there is an oily fattiness which appears when the craftsman laboriously strives for it. Therefore

it ascends from earth to heaven until it is
converted and becomes a bubbly fat, and the
aforesaid mineral quality is preserved in the
bubbles.

According to Plato, however, this operation is
not a natural one, for nature works to make
composite things from simple ones, because composite
elements are made from simple elements; but if it is
an operation carried out by the will, which is a
simple power that makes simple things of composites,
which is the same as producing an infinite power
from a finite one. So says Plato in his book "*De
quartis*."

If however, it is objected that the true nature
of things makes it impossible to turn gold into a
vapor by such operations, Plato says in the same
place that if one cannot make fire, that is, a
perfect simple substance, one may make air. If one
cannot make a circle, one may make a square, which
is only to say that simpler substances either are
made, or can possibly be made.

Experiment makes it plain that it acquires its
qualities from such an operation, for you may
dissolve an ounce of gold, once prepared, in a pound
of any spirit, and it fixes it in a day. If you
reiterate four or ten times, it fixes a pound in an
hour. A man may do many wonderful things in this

business, and we shall treat it as a philosophical question.

THE THIRD PART

As to the third question, how that power is implanted, you must know that a mineral earth is the more favourable, because we are made of such and are fed and nourished by them, as it says in the book "*De anima*":- But that (necessary for this task) is a mercurial power, therefore it ought to be sown in a mercurial earth.

Note, however, that the earth must first be cleansed of its thorns and excrescences and sublimed with a fairly strong fire, but nevertheless its central core must be preserved; to do this and save it from being burnt, a slow and gentle fire is employed. Besides, such a fire conserves moisture and perfects fusibility.

Two operations are found necessary on this earth or humidity, for the aforesaid power is nourished; as Hermes says; The earth is its nurse, that is, a sublimed and moist mercury, or something able to penetrate mercurial spirits; it is something like this that fills the *ampullae* of the aforesaid power.

He adds immediately that the wind, i.e., the water, carries that very power in its womb, i.e., by means of that power's womb. The reason for this is that its light parts cannot be choked down and prevented from shattering its earth in a spirit of rage when it finds itself squeezed by cold. So that power which is called wind waits to receive its full nature from the strengthening heat of the fire when it is shut up in an ampulla, especially if the quantity of unfixed material is in any way excessive.

However, if the reproductive power is strong enough to hold those spirits in, it fixes, and immediately converts things to its own nature, for this is how the power that the art is seeking multiplies and grows; but if the lawless fire holds sway for any length of time they fly away at once and are expelled. Therefore all should be again dissolved and again roasted until additions no longer cause ferment and change. As Plato says in "*De quartis*", The fixed fixes and the coagulated coagulates.

It espouses itself and inseminates itself; there is nothing in the world more wonderful. So Plato, *loc cit*, says that the brain is the seat of thought, or the thoughts of the mind, and the soul marries it because of its simplicity, according to (the work of Sol?). Therefore, he says, you ought to make the material as simple as possible, for it is then able to make use of simple materials and kill itself. So Maria, sister of the wise, said, Sow gold in leafy earth, not common gold, but philosophers' gold, which has been philosophically prepared, for it increases, grows and is fed like other plants.

So Aristotle confirms this, when he speaks to Alexander and says that the material is called the elementary stone, for the four elements, as previously said, are extracted from it. It is called the mineral stone because it is made from the minerals of the ground; and the vegetable stone because it is nourished and increased, which is the property of the vegetable soul; and the animal stone because it is made with an odour and broken up with a stench; and the rational stone because it remains in that natural state which was designed for it.

But a natural work is a work of divine intelligence and the material of the art arising from methods of natural occurrence and chance is here called by the name of all sorts of seed. When it is the animal stone it is called by the names of all sorts of flowers, fruits and liquids.

When you see the sign, that is, when it increases in amount, you may find out the secret, that is, the time when the mercurial substance retains its own nature, fusible, penetrating, entering, and performing all parts of the work that are sudden and complete, for it fixes by the power of its spirits, and remains moist and therefore able to perform fusion, entering and fixation.

But since not all sperm begets, it should be assimilated and digested in nutritive spirit which is generated in the vital liver and becomes an animal spirit in the heart, and the longer those spirits are digested and revolved in the places mentioned, the readier they are to beget and the more perfect the effect to which they give birth, and the more their number is multiplied; sometimes two, sometimes seven and sometimes forty are made at a stroke.

Avicenna says that he himself saw a pregnant woman abort forty fetuses at once; the philosophers have considered that the seed does indeed enter and propagate, but the aforesaid spirits by themselves, as is clear from the philosopher's book, "*De Animalibus*", begin the work closed in a glass vessel, and dissolved in water by use of (the heat of) dung, and afterwards are crystallized by a warm moist fire. If that vessel is opened, the substance is shattered, and the spirits vanish at once. They say that this should be repeated often, for according to Rhasis, solution and re-solution are the heart of the work, and its happy completion, and in that lies the whole secret.

For, as we said previously, the stone multiplies endlessly in quantity; as it says: At first one part will perfect ten; if dissolved and crystallized once, it perfects a hundred parts, if two or three times, two hundred parts, if twelve times then it perfects to infinity. Its nature is so powerful, as the philosophers have said, that some bodily or spiritual property of every creature is held and contained in the stone; such is the power of its effects.

When Hermes investigated it all, the power of the stone, and how it was of a higher grade than other natural things, he heard that it tried to reach the third heaven, but this was forbidden, for it is not right for any created thing to usurp the powers of its creator. So says Plato in "*De quartis*". In another part of the same book he says that solution takes place in Luna and crystallization in Saturn; therefore the stone acquires the powers of all the planets.

Likewise, solution takes place in water and crystallization on the fire; therefore it acquires the powers of the elements above and the elements below; so it shares in the signs and fixed stars, in moisture, and compositions and limbs, and winds, and regions of the world. Likewise because the flying spirit is like the angels and because the dead part becomes volatile, it seems to hide the office and mystery of the resurrection. Although all these things may be taken metaphorically, they say that it is of the greatest efficacy. So much for Plato.

Its effects are, so to speak, like those of the animal seed that enters the belly again and there reduplicates the nutritive spirit. The same may be said for the vital spirits of the heart and the animal spirits of the brain. It is certain that the power of the seed may be multiplied and increased many times; and if the material is administered in sufficient quantity, a living creature is made out of it distinct in its species and nature, as we said.

As to the fourth question, when men seek to gather the foetus before time, they procure an abortion, therefore philosophers ought to await the proper time. So that it can be better perfected within its own boundaries, I shall clearly state the infallible signs of completion.

They are such as we have previously mentioned, for it is thinner than air and whiter than than milk when it is at the white stage, and more glistening than red lead at the red stage. The white stone differs from the red stone only in the addition of a citrine color, which it can receive only from a Mercury more liquid than the element, more bubbly than fine foam, more spiritual than a fierce wind, clearer than aqua viva, which is thickened by the attacks of fire, but which remains totally uncrystallizable by any cold however great or any heat however slight.

Careful sublimation and frequent separation through the filter produces the first sign of the white stage. The sharp point of calcining waters produces the sign of the red stage.

Repeated solution in sharp waters with successive roastings produces the second sign.

Solution as in the bed of a stream and gentle crystallization produce the third sign.

A careful purification, cleansing everything, produces the fourth sign.

A fire that causes strong precipitation gives the fifth sign.

The sixth sign destroys the nature of the previous ones; besides all, the others are external signs of its great value. As they say, if you mix a grain of the elixir in clear wine it cures leprosy, scabies and impetigo, it immediately carries away all fevers and all heat, and removes all morbid moistures from the human body; it straightens every shrunken member and preserves youth. What else? No one who uses this as food or drink shall know infirmity at any time. He shall always be red-cheeked and cheerful.

There are other signs by which to recognize it. If you convert forty pounds of white or red mercury into water, and allow it to fume over a small fire, and project one ounce of the aforesaid elixir upon it, it converts it all into its own fixed state, and ferments it, and will communicate its own powers equally to it, and will cause all the aforesaid signs to appear in that water.

So also it makes crystal or glass fusible; it is malleable and confers malleability. Likewise it will convert any precious stone to any color you like, for it receives all colors, such as red; it unifies with and hardens any body it is mixed with.

As to the fifth question, one must say that the inherent quality of that active power is the flowing quality of mercury, but it must be purified from its stinking earth for the nature of a solvent is to produce fine matter - and it ejects its coarse part when it is purified, and when heated it afterwards overcomes and immediately pierces itself, and in piercing it digests and hardens; thus it produces the qualities of gold in an hour, as the wise have said.

Geber is witness that the portion of it that fixed what the fire previously carried away is conserved; it renders malleable because it has, acquired the nature of mercury; above all it is perfect when tested by fire, for only mercury has the ability to overcome and lie quiet on the fire - and some others.

Experiment proves that it hardens immediately, for the fumes coming off from Jupiter or Saturn harden it at once; and so any power which is infinite is more perfect. In the same way it has all the medicinal properties of gold.

To learn about this, one must first know how gold strengthens. Some have said that it does not strengthen by nourishing, like bread which strengthens the heart of man, for it is neither digested nor changed, nor does it Loosen or bind like this food. It does not comfort the heart by modifying odours and spirits, like perfumes, nor by purging excess, like laxative medicines. This is confirmed by the words of Dioscorides, who says that if any phlegm or moisture is weakening the spirits, gold wipes it away. But its cleansing power comes from its mercurial nature, as Avicenna says; so it can strengthen the heart in this way.

Others say that gold comforts the heart by its qualities, for gold is hard and the hard is strengthened by the hard, which is a saying of the great doctors, and Vincentius says in "*In Speculis*" that the harder something is, the better it works. Others have said that gold comforts the heart by separation from all its material nature, as a magnet attracts separated iron; but gold does it by its elements, that is the qualities of mercury and sulphur, which are the two basic elements of this natural species - and which, one may suppose, differ from each other as a man differs from a Franciscan; and so the powers of the whole species follow from the powers of its basic elements; moreover, it shares equally in what preceded it by reason of the powers instilled by heaven, and the heavenly powers themselves; and it flows into the composition of its passive complement, always and everywhere. Therefore a man generates another man, and so Sol, with its substance excellently arranged, and with more than natural power, will communicate such power. Therefore it is the same as it was before.

Others argue that it is pearls that strengthen the heart, therefore it does not abide by its own kind entirely. But one may reply that the heart is strengthened by special means, which are found only in gold. However, it is argued to the contrary that sapphire strengthens the heart by its own nature and therefore by special means. There are therefore special means that are special to both.

But pearls strengthen the heart and so does coral and many other things that are found in the world, therefore there is an infinite number of special ways. Therefore gold can indeed be prepared in certain ways and by fixed intermediaries, as was previously said. Thus the question is resolved and the true and certain answer given as regards the nature of Sol and Luna. Praise and glory to God through the ages forever. AMEN.

PROCESSUS SINGULARIS DE MATERIA CHAOTICA

By

Martini de Delle

A man from Milan, born in the town of Vitry.

A true and venerable Adept, and an F.R.C.

Translated from the German tract:

"Geheimnisse Einiger Philosophern und Adepten, aus der Verlassenschaft eines Alten Mannes" edited by Chris. Hilscherr 1780

By: Léone Muller

At the very first storm of the year, collect the rain which falls from heaven, before it touches the earth, in big retorts, lute them well, set them in the sun to putrefy the water, and leave the glasses there until worms begin to grow in them. Then distill it until it becomes as clear as crystal. Thereafter, pour the water into clean phials, lute them well and keep them in a cold place.

On precisely this day, after the storm, when the sky is clear, (about dawn), you should also dig a hole out in the open air in a ditch of white or red clay, or loam. Thereupon, moisten the sides of

155

the hole with the urine of a boy and immediately cover its opening with straw and leaves. Do not cover it TOO thickly but merely to prevent anything impure from falling into it. Permit the sun to shine upon it during the day and the moon at night; also allow the air to penetrate it. The following morning you must return to the hole before sunrise and you will see some white salts on its walls. Carefully gather this into a clean phial (vessel) with a glass rod or spoon. Continue doing this until you have a good quantity of this. Now seal the phial and keep it in a cool place.

The salt which you have thus obtained is dissolved with the preserved, distilled, thunder water which has been filtered. The moisture is allowed to evaporate again to a salt in evaporation bowls or dishes. The dissolutions and evaporations are continued until no more feces are left at the bottom of the dish.

Thereafter, put this pure salt into a high glass bowl, cover it well, and bury it in Ehe earth. The salt will resolve into a whey-like water which will attract to itself the power of heaven and earth. Following this, pour this thick water into a clean phial, lute it, and put it temporarily in a cold place.

Now put part of the *purpura Solis* or *crocus Solis*[8] into a phial. Pour upon it six (6) parts of the wheyish water, immediately lute a blind alembic on the opening and set the glass in a vapour bath in the first degree of the lamp-fire. In so doing, make sure that the lamp never goes out until the matter has gone from blackness to whiteness. After this, remove the glass from the Balneum and set it in a slightly heated ash-cupel. Give the second degree of the <u>lamp-fire</u> until the desired end is achieved, and you will see an *eclipse* of the Sun (Gold), starting with a fire in a glass in a furnace. After that *Cauda Povonis*, (or: peacock's tail), and finally the supreme whiteness, and at the end, the most perfect redness. All this happens of its own, by a fire in a glass. *Quod sone Sophistus incredible videtur.*[9]

When now this matter has become dark red, you must take it out, weigh it, pulverize it on a marble, put it into another phial, pour upon it twice its weight of the wheyish water. Then, lute a blind alembic upon it, set the glass back in the vapour bath and continue to proceed as according to the prescription. This will cause the matter to multiply ten-fold each time. This work can be repeated. Finis.

---

[8] purple or yellow gold -HWN
[9] that which indeed seems incredible to the Sophists -HWN

ARCANUM ARCANORUM ARCANISSIMUM

Translated from the German tract:

"Geheimnisse Einiger Philosophen und Adepten, aus der Verlassenschaft eines Alten Mannes" edited by Chris. Hilscher, 1780

by: Léone Muller

ITS PROCESS

In the Elements two kinds of sperm are generated; which are specified from Christmas to the end of the month of May, partly in water, partly in earth. The first is called *Axunga Solis*, or *Mas*, because it contains much Sulphur and little Salt and Mercury. (*Axunngia* is rendered as 'axle grease, kidney suet or animal fat'). The other is called *Marga Lunae*, or *Foemina*, as it contains much Salt and Mercury but little Sulphur.[10]

The Axunga Solis arises each time from morning to noon and, at sundown, falls on the water as a white scum mixed with many colors which the wind afterwards drives to the curves of the banks. At dawn it is gathered, if possible cleanly, at the rate of 50 or even 100 lbs., with a net or other instrument into clean big sugar-glasses (which must

---

[10] Marga is rendered as 'marrow or marl or a kind of earth'. May mean: "Solar and Lunar seeds". -HWN

not be filled above one-fourth) and kept tightly protected from the sun and the air.

This Axunga Solis is also found before sunrise in meadows and pastures. The very purest, however, is found on mountains and looks like a white and green jelly or thick frog-spawn. But when there has been dew or rain in the night, it rises like a blown-up bubble and turns black and seeps into the earth as soon as the sun shines on it.

The Marga Lunae, on the contrary, comes from the North and is found before sunrise, from Christmas to the end of May, in meadows, on trees, and on the mountains. It is oily and looks like white intestines mixed with many colors, simultaneously quite pure and heavy. It contains much Salt and Mercury but little Sulphur. Collect of it likewise 50 lbs. and preserve it from the air and the sun in the shade in clean glasses filled only to one quarter (but the glasses must be well closed).

If now you wish to prepare the Tinctura universalissima (Supreme Universal Tincture) from these two substances, take one part of the Axunga Solis and two parts of the Marga Lunae, put them into large sugar-glasses which must be filled with one part of the matter only, leaving three parts empty. Otherwise the glass would be shattered. Thereafter close the glasses with double wax paper and set them in the sun in the spring and summer till the fall, but always take them into the room at night. In the winter leave the glasses in a warm spot in the room, and the matter will process itself within one or two months. The Nature-spirit will warm up and cause it to putrefy and this without any external fire. When the matter has turned black, the blackness will disappear again in due course, and the matter will be transformed into a red liquid, which will amaze you. Let the red liquor stand till the Earth has separated from it and has settled at the bottom. Then the first purification is completed.

Now pour this red viscous water into other sugar-glasses to allow the Earth to separate completely from it (but guard against the smell of this black earth as it is quite poisonous and attacks the lungs). Keep it separate, then close the glasses again tightly and let them stand in a warm spot or in the sun. The Nature-spirit will again begin to work of its own, and on the water a thick scummy skin will grow, upon which Mercury will sprout like the finest pearls. With this another purification has occurred.

Now have ready another clean glass, remove the white skin with a glass spoon and put it into it, then close it.

After this, close the glasses with the red water once more tightly and let them stand undisturbed till a white skin has again formed on top. Add this to the first skin in the previous glass, and continue doing this till the red water no longer produces a White skin. Then you have accomplished the third purification.

The thus gathered white skin or Mercury again goes into putrefaction and turns black, but when it has stood for some time, it is transformed into a red water upon which you will find a yet much purer skin.

Remove this skin each time and preserve it in a new glass till the red water no longer produces a skin. Then the fourth separation has come to an end.

Now let the glass filled with the gathered skin of Mercury stand quietly in a warm spot, and it will be resolved into a golden-yellow water, dropping a white Virgin Earth. On this water, however, you will see the Double Philosophical Mercury like a pearl-tree. Gather it carefully into a new glass. Preserve the golden-yellow water and the White Earth also very specially as a great treasure till their further elaboration, and the Nature-spirit has completed the fifth separation.

NB. If you now wish to prepare the Supreme Universal Tincture with this golden-yellow water or Double Spirit of Mercury:

Put one Quentgen (1/4 Lot or Loth) to one Lot (1/2 ounce) of Solar calx egr., reverberated to the utmost, into a phial closed with a ground stopper, the thickness of a thumb. Gradually add to it by drops 16 to 20 Quentgen or Lots, according to how much you have got. Now screw the phial tight and seal it with the secret Hermetic Seal or mass, put the phial in a quiet place, and the gold will first dissolve into a green, then a red, then a dark red, but finally a black liquor. Thereupon the blackness will again disappear and the liquor will turn white, yes, like milk, at the same time dropping its White Earth. Now various colors will appear, which no painter could produce more beautifully.

(N.B. This golden-yellow water is a penetrating spirit, tasting like sugar on the tongue while simultaneously being a great fire.)

But continue letting the phial stand immobile, and the liquor will again become red and look like thick blood. Over the coagulated matter a brilliant star will arise. Following this, the matter will become so spiritual that it will break the glass. Now our Work has reached the desired end, and by His Grace God Almighty has granted you the Universal

Tincture. It has produced itself by the inner fire of Nature, and thus without any external fire, in and by itself, and you had no work to do yourself.

True, it is a slow work which often leads into the 4th., yes, the 5th year, but you cannot wonder enough how the inner fire operates in this matter, cooks it and brings it to the highest perfection of the Universal Tincture. But in this Work you must take careful heed not to disturb this pure Spirit in its operations, and then it will gradually bring everything to the desired end.

If now you wish to test your heavenly tincture to see what powers it has in the transmutation of base metals, take 3 lbs. of a well purified Mercury or a simple antimony Regulus. Put it in a crucible over the fire, and when it begins to fume, throw on it one Quentgen of your tincture, wrapped in wax, which will be followed by a bang and a thump. You must therefore use a large crucible to prevent the Mercury from getting out. Thereafter increase the fire, and after 2 hours the Mercury will have melted into a red salt and incomprehensible tincture.

After this, take 16 Lots of the finest gold, put 1 Lot of your red salt in flux on it, and the gold will again turn into a tincture.

Of this last tincture put 1 Lot on 1000 Lot of other metals in flux, Saturn or Mercury, and they will be transmuted into 24 carat gold.

CONTINUATION OF THE WORK

If you wish augment the Spirit of Mercury or the golden-yellow water which you had separated from the White Earth, do as follows.

Take 1 Lot of gold-leaf, purified to the utmost, and 4 Lots of live Mercury of native cinnabar, or antimony. Make an amalgam of them, wash from it all blackness, dry it, then put it into the Spirit of Mercury and seal the phial. Let it stand undisturbed, and you will see wonder upon wonder, as the Spirit of Mercury will open the gold and transform it into a green liquor. In it everything in the phial will grow, such as trees and precious stones with intermixed gold and silver grains. This will soon disappear and look quite black, and out of that blackness little fire flames will thereafter leap up.

After the blackness has gone, the most beautiful colors will appear. They too, will disappear, and in the phial something like a green meadow will present itself with various kinds of precious stones which likewise change into a crystal, looking like candied sugar, that is, with intermixed yellow colors. After this, it looks as if everything were to become green velvet. Then it changes to a yellow color and finally to a red one which, however, will continue to become brighter. At last, a really blood-red and brilliant substance is left, and on it will appear a purple star. That then is a sure sign that the matter has been congealed into a fixed and liquid salt which is the greatest mystery of all Nature, with which no thing on earth can be compared. With this salt or tincture you can proceed further, as has already been taught.

But if you wish to proceed in yet another way, take the Spirit of Mercury with its white Virgin Earth, put them together into a small retort, set it in the sand with a receptacle, and the Spirit will go over together with the Earth, unchanged, sweet as sugar and burning like fire. Now make an amalgam with gold and Mercury, as already described. Put it in a long-necked phial and pour on it the Spirit together with the white Earth that has gone over with it (but keep the Earth), seal the phial, set it in the B.M., in moderate heat for 14 days and nights. Everything will open into a green liquor which decant into another small retort, lute a receptacle on it, and put the retort in sand in gentle heat. Everything will go over except the Earth of the gold. Now add to it the Virgin Earth which you had preserved.

166

At this stage there are again 2 ways ahead of you: In one you must divide the liquor or Spirit into three parts and let the first go through the colors in a sealed phial, while augmenting the tincture with the other two parts.

In the other way, however, you must divide the Spirit of Mercury or the liquor into two parts, put one in a flask and set the flask in a B.M., distill the volatile spirit 7 times from it, while pouring it back on each time. Also, cohobate it 7 times with ashes and 7 times with sand. You will obtain only a little Phlegma. Finally, let the flask also stand for 4 days and nights in iron filings till everything has become fixed. Thereafter remove the tincture and put it into a retort. Then again divide the other part of the liquor into 2 parts, pour one part on the tincture of the retort and drive the Spirit of Mercury over in the prescribed manner, making it fixed again. Proceed likewise with the third part of the liquor till everything is fixed, and this tincture will tinge like the first

Here now you can see what kind of a secret God Almighty has enclosed in the Celestial Sperm, provided the Nature spirit is not hindered in its operations, because it operates without stop, and there is nothing in the world that could be used to assist in this Work - because the matter contains everything in abundance. It is the cinnabar ore of the ancients, because it gives off a blood-red solution which separates its Mercury, and no artist can accomplish this you must leave it solely to the operating Spirit of Nature.

IF NOW YOU WOULD ALSO LIKE TO SEE
THE GREAT SECRET OF CREATION:

Take 1 lb. of fresh May dew, let it stand for 40 days and nights to allow all impurities to settle. Then filter it and have at hand a double glass sphere, whose parts must fit precisely in the center. In the upper half there must be a small opening. After this, take your pure dew, put it into the sphere and close the sphere with the secret hermetic seal, or our secret mass. When it has hardened, throw into the small opening of the upper half a grain of rapeseed of the tincture prepared with the astral sperm. Screw the sphere tight with a ground stopper and leave it undisturbed in a place that is not too hot or too cold. The first day, the dew in the sphere will become black and everything will be dark. The second day, open the sphere and do not hesitate to throw in yet another grain of mustard-seed, and the dew will become as clear as a crystal but with some darkness appearing in the depth or at the bottom, and you will see the World Spirit move in its light.

The third day, again throw the weight of a grain of mustard-seed into the sphere, and the World Spirit will move everything together inside it. At last, however, the separation will take place and push the fiery Earth to the bottom. Above this fiery Earth there will take place and push the fiery Earth to the bottom. Above this fiery Earth there will stand a milky viscous water, and above that a clear, crystalline blue water with golden-yellow drops of oil, showing that the Earth underneath has been hardened by the fire of Nature. That fire also dwells in the Earth and attracts the upper fire. The fact, however, that a milky viscous water stands on this Earth is the sign that the Earth is thereby preserved, so that it will not ignite itself and burn. It is out of this water that the Earth draws its nourishment, and in this rnilk there is the food for all creatures, and their power of multiplication, yes, it consists of Fire, Air, Water, Earth as well as of the spiritual viscous Elements, as Hermes says:

Ab Aere quasi in utero gestatur, nutrique a terra.[11]

The golden-yellow oil-drops on the azure-blue crystalline water signify the sun, the moon, and the

---

[11] It gestates in air as in a womb, nourish it with earth. -HWN

stars, which burn and sparkle through the light-spirit, while the water signifies heaven.

IN CONCLUSON

I will now reveal to you the true *Sigillim Hermetis*, supreme secret, with which you must not only seal the glasses but also reopen them. In general, there are still much greater secrets hidden in it, but which the Philosophers were very careful not to divulge and considered worthier than even the Philosophers' Stone.

Make an amalgam of 3 Lots of English tin, beaten into fine leaves, with 5 Lots of pure live Mercury in a glass bowl. When it has turned into an amalgam through trituration, mix it with 8 Lots of unadulterated sublimated Mercury, taking care not to add either gold silver, or any other metal. Thereupon put it into a retort with a wooden scoop.

Or: Of the beaten English tin and live Mercury take 1 lb. in equal parts, pound it in a glass mortar into an amalgam, mix 2 lbs., of sublimated Mercury with it, and put the powder in a glass retort with a big receptacle into which you must previously have poured 3 Lots of the volatile Spirit of Mercury, that is, *Spiritus Roris Majalis*, (Spirit of dew) the processing of which will follow at the end. After you have well luted the receptacle to the retort, set it in iron filings and melt them in the beginning with a very gentle fire, and a clear spirit will go over with great impetuousness. If you give too strong a fire, however, the spirit will break the receptacle into a 1000 pieces. It is therefore necessary to put the receptacle into cold water, to allow the all too strong heat to cool down in it, as in this way everything goes well. Finally, increase the fire to make the retort red-hot. Then let the fire go out and the retort cool.

With the spirit that has gone over mix the sublimate which you have received at the same time. Pour them together into another retort, lute a receptacle quite tightly fitting on to it as this spirit, like the Nature-spirit, at once begins to coagulate, which should not yet happen at this time. Now put the retort in a sand cupel, distill the spirit over once more, and you will obtain a crystalline liquor. Pour this into a new phial, seal it, and put it continually in a warm place.

This spirit makes all glasses soft and tough, so that they can be worked with one's hands. The glass itself in which it stands becomes like a leather purse which can be pressed together and opened again. But as soon as this spirit is put in a cold place, it coagulates into crystalline butter, and it is impossible to remove it from the glass.

That is why the glass must be kept standing in a gentle heat.

If now you wish to prepare the crystalline matter or our secret seal, take some transparent crystalline pebbles that are not streaked with green, red, or yellow little veins. Make them red-hot in a crucible, then quench them in a stone vessel that contains some of the Universal Spirit, that is, the *roris majalis*, (the May Dew), and they will be reduced to a white powder. Pound this in a glass mortar to a very fine consistency. NB. Take care, however, not to add anything metallic to it. Of this powder put 2 Lots in a double glass bowl and pour 4 Lots of your liquor on it. Keep what is left over in gentle heat. Now the powder is dissolved and everything is transformed into a crystalline substance, together with the glass, and everything will be shining and as soft as butter, just like the substance in the glass. You must also keep this matter, together with the bowls, on a continual heat, for as soon as the substance has contact with the air, it becomes as hard as a diamond.

NOTE. Another Philosopher teaches that one should pour as much of this spirit over 12 Lots of finely pounded white glass in a burnt stone pot as will make a pap of it. Thereafter it is to be put in gentle heat in ashes, well covered. The pulverized glass will become like a resin within four days, by which anything poured over it will turn hard like a diamond in the air.

When now you wish to seal your phial, rub the glass stopper with it. Then take some of it from the bowl with a clean piece of wood and form it with your hands, for as soon as the matter reaches the air, it turns into a crystalline dough. Take this out and put it around the grooves of the stopper, and it will close everything and become harder than the glass itself. This is the secret Hermetic Seal of the wise, because with it they can perform miracles, and it is to be valued much more than the Philosopher's Stone.

If now you wish to detach the seal from the phial, coat it twice with the double liquor which you must at all times keep in a warm place. The former substance will again become soft, so that you can remove it. You must keep this in a clean glass, in a warm spot, for further use, because only a little of the double spirit is added each time.

But so that you may learn that this seal is to be valued more than the Philosopher's Stone, I will teach you in detail in the following chapter. Only of this I will here remind you: If you use some of our diamond substance for sealing, you must always use a fresh supply, because it becomes contaminated even if you keep your hands ever so clean. Some dirt will adhere to it, so that it cannot be used again for sealing. This is truly the greatest mystery.

HOW TO PREPARE VARIOUS DIAMONDS AND PRECIOUS STONES FROM OUR CRYSTALLINE MATTER, WHICH ARE FAR MORE PRECIOUS THAN THE ORIENTAL ONES.

This then is done as follows: When you are preparing the abovementioned crystalline substance, you can form it into various figures, but you must take care that no impurity gets mixed up with it.

The forms must be of gold. If now you wish to shape a figure, let it lie outside in the open air for one night, and it will become harder than a diamond. Thereafter put it in a clean glass and pour enough of the seven-times rectified Universal Spirit over it to cover it completely. Now close the flask with a blind alembic and set it for one month in gentle heat, when the diamond or the figure will get a great brilliance. Then take it out.

But if you wish to prepare a ruby of it, take some of the very finest gold and have it beaten into thin little leaves. Dissolve it in our Universal Menstruum, as you heard before, and you will obtain a grass-green solution together with the separation of the Earth. Take 1 Lot of this gold solution and 2 Lots of the prepared pebble powder, put it in a round grass sphere and mix it with 4 Lots of the glass-making substance. Screw the sphere tight and set it in ashes in gentle heat for one month. You will get a fiery transparent substance. Now let it harden in the air, as has already been taught. After this, put the substance in a sphere and pour some of the seven-times rectified Universal Spirit upon it, as much as has already been mentioned. Let it rest in it for one month when the ruby will develop a mighty light, causing it to shine at night like lightning. You can have the required form prepared like a star, together with the cover of gold, but they must first be hollow-ground inside.

If you wish to prepare an emerald, however: Take some Japanese copper and dissolve it in the universal Menstruum or Spirit. The copper will be colored grass-green. Now separate the solution from the feces and add to it 1 Lot of a gold solution. Proceed with this as has already been taught, but better use its greenspan. (verdigris)

If you wish to prepare a ruby: Make a crocus of iron, as I have taught in my writings. Take of it one Lot and pour it on 4 Lots of the universal Menstruum or spirit. Extract it in gentle heat, and the Menstruum will turn red. Pour this off into another glass. Mix one Lot of this with 2 Lots of prepared pebble powder and 4 Lots of crystalline substance - but not with your bare hands or something metallic. Use a clean piece of wood. Do it with the greatest speed in a warm spot, (NB. This must be done in connection with all stones) and proceed in everything as has already been taught, and you will obtain a ruby of inestimable worth.

BUT IF YOU WANT A TURQUOISE THAT SURPASSES THE ORIENTAL ONES:

Take one Lot of pure silver, beaten into leaves. Pour 4 Lots of the Universal Menstruum on it in a flask, set it in gentle heat, and the Menstruum will turn blue-green. If the turquoise is not to be transparent as the oriental ones are, take one Lot of that solution but previously well shaken, so that the Virgin Earth that has settled at the bottom becomes part of it.

If it is to be transparent, you must take one Lot of the pure solution and proceed as you have been taught in connection with the preparation of all stones.

But if you wish to prepare an amethyst: Take 1/2 Lot of the solution of the crocus of iron and

1/2 Lot of the silver solution, in the previous composition, and proceed as has already been taught. You will obtain a pure amethyst. And in this way you can make all stones

Lead results in the most beautiful carnelians (a variety of chalcedony) tin, instead, in white, milky, colored stones mixed with rainbows. Thus you can see what the Light of Nature is capable of doing after it has been separated and, as a primordial Spirit, reveal the *Mysterium Magnum*, (the Great Secret) in all things. In this way I have also revealed to you the Universal Chaos, the foundation out of which everything follows according to the Order of Nature.

Because in the preceding pages the use of a Universal Menstruum is required, I will reveal to you its preparation from the book of Trithemius, "*The Light of Nature*", Book I, Chapters 2 and 3.

From dew and rainwater collect an <u>Ohme</u> (an old measure for wine, a cask containing about 150 Liters or quarts) full of dew in the first months of the year, or of rainwater that has come down with thunder and lightning.

NB. The Universal Spirit is mostly present in dew.

Let it stand and putrefy of itself for one month, then stir it well together and first distill the most volatile and spiritual part off in a flask. Rectify it 7 times, then keep it in a big spacious glass, as the spirit would otherwise smash the glass into pieces, after which leave the glass in a cool place.

Thereafter, distill all the Phlegma and keep it for the cleaning of the salts

When everything has been distilled to the consistency of liquid honey, let it cool and pour this honey-like residue into a retort, put it in an ash cupel and distill the Phlegma over with the azoth and the golden-yellow oil. Separate this oil in a tritorium and keep it in a clean ground glass. After this, take what is left out of the retort, pound it small, put it in a flask, pour the azoth with the distilled Phlegma back upon it, digest them together, then decant and filter to remove the few feces. Now add to it half of the spirit which you had kept, set it in the B. M., with gentle heat then distill the Spirit off. When everything has cooled down, pour the Spirit back upon it and set the flask in the ash, distill the Spirit off again, and when big drops appear, quickly add another receptacle.

Distill the Phlegma over, and when acid drops appear, let the fire go out. When everything has cooled down, pour the Spirit back upon it once more, set the flask in the B. M., and distill half of it off. Put what is left in a cool place, and a heavenly salt will sprout. Collect it carefully. What does not sprout must be evaporated to salt. Purify the latter with the phlegmate by dissolving and coagulating it till it flows in the fire like oil. Now pound your first crystal small and add it to the salt which has been made to melt in the fire like oil. When the salts have cooled down, add the other half of the Spirit - but the latter must have been rectified 7 times to transform it first into a quite volatile and pure spirit.

Set the flask with the matter in the B. M., for 2 nights and days. Then pour everything together into a retort, set that in ashes with a receptacle and distill half of it over. When everything has cooled, pour back what has been distilled over, and in this way distill half of it twice more. When everything is cold, pour the spirit back once more, set the retort in the sand, distill half the Spirit over, also cohobating 3 times.

When everything is cold again, pour the distilled Spirit back, put the retort in iron filings and distill over *per gradus*, the last time with a strong fire, and everything will go over as a Universal Menstruum.

Pour this Menstruum into another retort and distill it twice more out of the sand. Now it is perfectly prepared for radically dissolving all metals and separating the Earth from them, so that the Philosopher's Stone can afterwards be prepared. Details can be read in Chap. III. of the Nature-Light of *Trithemius*.

This is the Universal Menstruum of the world, prepared out of the Universal Chaos, the key for opening everything in Nature. It separates the dark Earth, so that the sweet fire and Light can thereafter be obtained in pure form. This blessed **Maym**, is indeed the blessing of the great Spirit of Jehovah, by which He moistens the land and renders it fertile. It is the Fire of Life.

But if you wish to try another experiment with this Spirit, take some Regulus of antimony of Mars, melt it together with one Lot of fine gold, then pulverize it, and pour the powder into a big phial, again pour as much Spirit over it as there has been powder, lute the opening and set the glass in a mild digestion which will extract the Spirit from this powder. Pour that into another clean glass and keep it. You must repeat 7 times the pouring on of fresh Spirit in the amount described, in addition to digestion and decanting. Thereafter pour the extractions together and abstract them several times. Then you will obtain a beautiful salt after the last abstraction and refrigeration, like a diamond, which will have wonderful powers in medicine.

SAPIENTI SATIS

A UNIQUE PROCESS FROM PRIMAL

MATERIAL

In the months of March, April and May, three days before the full Moon, the solar Sulphur ascends from the Earth because of the heat, and exhaling like a white smoke to the higher regions of the Air, becomes totally spiritual; for the lunar cold raises it at this season. In the morning, therefore, when the rising Sun strikes cold Heaven with its rays, the frozen Moon turns the weighty solar seed and the moisture of the Earth into a clear Water, in which the four elements liquefy together with the seed of gold.

However, as soon as the Sun hides his brilliance, it (the solar sulphur) is again drawn down, because it has begotten that (water) which is close to the Hermetic substance, the Sun its father, the Moon its mother, its nurse the Earth, and the Wind carried it in its belly; Then that water is taken and after philosophic sublimation it is throughly purified in the Hermetic Vase, the furnace and the fire, as was seen. Many have possessed this genuine material, but unless they knew the philosophical manipulation, Nature would not produce the living Sulphur of Gold from the Earth, where it naturally follows that everything is done, is born, and appears, in a wholly philosophic fashion.

After you have collected that Water of May dew, at the proper time, as described, it is filtered several times; then pour the depth of two fingers, crossways, into a large glass, which is then tightly tied in a sheet of calf's bladder, and hung up in the heat of the Sun and cold of the Moon for a whole month, so that they are soaked in the bright burning mirror of the sunshine, and the Sun dew becomes impregnated with its inward strength, until at last it acquires a golden colour.

Now at first the water draws down blackness and stench, which disappear little by little, then it becomes yellow and acidic like vinegar, then this sign will disappear; shortly afterwards it becomes green and a fat resembling May butter is seen on the surface of the water; again this sign vanishes and at length a green Earth is produced, which rises till it ripens, then remains at the bottom of the glass.

Then the living Spirits will rise to the top of the glass and stick together above like the eyes of living fishes; but they descend to the Earth again and in a few days the living spiritual Sulphur of Gold emerges from an earth of the same colour and it will shine from its Water like a burning coal, beyond comparison with gold or the shining Moon.

Then it is ripe and ready to be distilled. Distillation should be done over a water-bath, and a third of the water should be distilled off. The remainder should be taken separately and circulated over the liquor or Earth by distillation ten times, continually pouring it back again. (Cohobation & distillation).

The earth finally remaining in the bottom of the glass is weighed, and moistened with a finger's breadth of the rectified Water, then placed on a gentle heat, and again dried; this is to be repeated until the Earth has imbibed three times its own weight, or weighs three times more than it did before.

When this has been done, the Earth is distilled in a retort over a strong fire. Then the celestial Mercury passes over into the receiving vessel, and is pure Living Gold, which puts all manner of diseases to flight, and renews the powers of man.

Coagulation of this red oil gives the Salt of Saturn, which should give a red dust when it has been fermented with gold. When the Stone has finally been received by gold, then it should be projected on all the other metals, as has already been described in the preceding processes.

Or this: Take water in which our Stone is, as you find it, which is like nothing else. Take nothing away from it and add nothing to it, for it must be prepared according to its own proper nature. Place it in a cucurbit and separate. Take the moist from the dry, and the body alone will remain in the glass, and the water will go over into the receiving vessel; marry this again, as you know how, and thus the fixed will become volatile and the volatile fixed. What was dead lives, and what was alive is put to death. Thence prepare the Medicine, which changes from colour to colour.

Finis.

THE DIVINE MAGISTERIAL SALT WHICH TINGES ALL METALS INTO GOLD

By the most Illustrious American Prince:

Holdazob A. Dachem

Translated from the German tract:

"Geheimnisse Einiger Philosophen und Adepten, aus der Verlassenschaft eines Alten Mannes" edited by: C. Hilscher, 1780.

by Léone Muller

Take 1 Quentlein (1/4 Lot) of gold poured 7 times through antimony, likewise 1/2 Lot of the very finest silver, 1 Lot of the best bismuth ore. First let the gold and silver flow together. Then gradually add 2 Lots of the pulverized ore. When they are flowing together like water, let the fire go down and the crucible cool, and you will find a Regulus.

Powder this and mix it with 2 Lots of pulverized oriental pearls. Put the powder in a retort, pour upon it some of the best philosophical spirit of vitriol q. s., till it has completely absorbed the powder. After this, put the retort in digestion and draw the spirit back again. At the bottom of the glass a calx of various colors will be left. Upon this calx pour enough distilled dew water to dissolve it completely. Now filter the solution and precipitate it with a rectified spirit of vitriol. Draw this spirit off again to the consistency of salt.

This Prince said that the virtues of this salt would be much greater if it were repeatedly dissolved with distilled dew, filtered and again evaporated to the consistency of salt, till no more feces were left.

With this salt the greatest medicine for human beings can thereafter be made as follows:

Take 3 grains of this salt, put them in a filter, pour upon them 2 Lots of distilled plantain water or 2 Lots of most highly rectified spirit of wine. When one of them has run through, pour once more 2 Lots of plantain water or spirit of wine into the filter. Repeat this work 7 times, and the water or spirit will look like blood. Preserve it carefully in a strong glass.

Of this filtered water, which will be 14 Lots, give a patient every day 2 Lots, and he will be relieved of his deep-seated illness with the seventh dose, no matter what it may be.

Of the spirit, instead, give 10 or 12 drops in water or wine, and you will see wonders.

If you dribble of this spirit one drop on a silver sheet, it will color it like the finest gold, NB. provided it has been hot.

But if you wish to prepare the Philosophers' Stone with this salt, proceed as follows:

Make it magnetic, that is, dissolve it after precipitation with spirit of vitriol - not with distilled dew. Have an evaporation bowl at hand and put in it half a finger's breadth of this salt. Spread it out and expose it to the rays of the sun in the morning to enable it to attract to itself the remaining Phlegma in the Salt. But you must begin with this work in hot summer days. Thereafter keep the salt Ln a strong glass, well luted, in a cold place.

When April or May arrives, pour the salt into a big evaporation bowl and set it at night in the open air, so that the moon may shine upon it. Then the salt will magnetically attract water from the air and melt thereby. In this water is contained and hidden the general vital food of the air.

The following morning set the bowl in the sun and let it evaporate again. The left-over salt will only have retained the seed of the sun and attracted to itself the Phlegma.

NB. But you must set the bowl in the sun every morning and keep it in the room at night, covered - do not set it in the air at night till the sunbeams have drawn off all the phlegma. The philosophers call this coagulation a philosophical day.

You must repeat these dissolutions and coagulations several hundred times, till the salt no longer attracts any moisture in the night but stays dry.

When you wish to test this salt, put some of it on a red-hot copper sheet. If it melts on it and does not smoke but tinges the sheet into gold, laud and praise God, because you have obtained the Philosophers' Stone, that ls, when you are so fortunate as to have perfected this salt from the month of April till the end of July or August.

Finally, remove your magnetic salt from the bowl and put it in a strong phial of which 3 parts must stay empty. Seal it hermetically and from June to the end of August set the glass in the sun during the day and in the moonlight at night to prevent either rain or thunder from touching the glass. Then all the colors will appear in the glass until at last the beautiful redness emerges, which is the greatest treasure in the world.

If now you wish to transmute silver into gold with this red salt, proceed as follows:

Put in an evaporation bowl 1000 parts of calx of silver that has been precipitated with salt water and has often been edulcorated. In the middle of the calx put one part of your red salt and set the bowl uncovered in the moonlight at night, again in the sun during the day, and the silver calx will dissolve into a water within a Philosophical day by virtue of the magnetic salt. Proceed with this water as I taught you before, and this calx will become a tincture.

Divide this tincture into 2 parts. With one part you can tinge tin or quicksilver into gold in the following way:

Put 1000 parts of quicksilver, tin, or lead in a crucible. When it is in flux, add one part of this tincture, wrapped in wax, and let it flow well covered for a quarter of an hour. Then pour it out and it will have been transformed into the finest gold.

With the other part of the tincture you can ferment the silver calx *ad infinitum* and change it back into a tincture if afterwards you again put one part of this fermented tincture on 1000 parts of

silver calx, proceeding as I have taught you.
L.D.I.S.S.A. (Praise be to God, etc.)

THE MAGNETIC SALT

With it tinge 1 grain 30730 drops of tap water as red as blood. Indeed, if in the months of May, June, or July the glass is put in the rays of the sun at noon, mixed with one grain of this salt, you will notice after two hours how the rays concentrate in the water, coloring it like blood. But as soon as even one single cloud passes before the sun, the redness will move to the center of the water, while the water will be bright and clean above and below. If the glass is left in the open air, however, a skin will appear on the surface of the water, but after sundown a black earth will have settled at the bottom, which can safely be left there for 8 days. Now pour the water off by inclination, dry the earth, put it wrapped in wax into flowing silver and refine it with lead. One grain of astral gold will be obtained after the separation in Aqua fort, for which God's omnipotence should be admired, and praise and thanks given.

NOW FOLLOWS THE PREPARATION OF THIS MAGNETIC SALT

Take 1 Lot of fine silver, let it flow and add to the flux 4 Lots of powdered bismuth ore. When both have well flown together (NB. But be on guard against the fumes, they are poisonous), let the crucible cool, pound the Regulus that you have obtained to a fine powder, put it in a flask, pour upon it one part of strong spirit of niter mixed with 3 parts of distilled vinegar, close the opening, set the glass in a mild digestion and shake it frequently. Then pour the solution in another glass, add again fresh Menstruum, proceed with imbibing, extracting and decanting till you taste no mere vitriolic sharpness. Thereupon pour all of the extractions together and keep then carefully. On the *Caput Mortum*, (Death's Head) pour some of the distilled dew water q.s., boil it thoroughly, filter the solution and add the solutions which you have preserved to the water. Evaporate it in a glass bowl *ad cuticulum*, and put the glass thereafter in a cold spot. Soon crystals will sprout. Dry these and keep them in a clean glass with a ground stopper, in a dry place, and you will have obtained the miraculous, magnetic salt.

ARCANTUM

This was in the possession of the Jesuit Father Nicolai. It was, however, taken from him by another Brother, called Joseph Antonius. When the latter was travelling through Leipzig and desired to die there,

I.N.N. obtained it from him *sub sacramento*, when he said to me that I could thereby become fortunate.

This process has been elaborated by His Imperial Majesty, the blessed late Emperor Francis, together with his director of conscience, Father Franz, and he left a large quantity of pearls and precious stones.

My Brother, I will hereby reveal a red earth, but which cannot be obtained more than once a year. You must understand the constellation of the planets, otherwise you will fail, because you cannot go by the common calendar. The best time is when the change occurs in the sky. In mountain towns it happens in the middle of June, at the full moon, for then the moon enters Cancer at 3 a.m., and in this hour the red earth looks like pure gold. It can also be obtained at Freyberg in Meissen, at Annaberg and at Claustal. This earth looks brown before the aforesaid time, but with the effect of the weather it looks like the finest gold. This color does not last for more than 3 or 4 hours, however. At the above-mentioned time several oak barrels must be filled with this earth and immediately shut tightly. Then return home with them. This earth contains the true Universal Spirit.

Thereafter only half of this earth is put in retorts, but in such a way that 2 parts of the space stay empty. Lute several big receptacles on, then set the retorts in sand cupels and at first give a very mild fire. A spirit will go over with many colors. When none goes over any more, the fire is allowed to go down and the retort to cool. After this, the spirit is poured into a strong glass and carefully preserved. The salt from the caput mortuum, however, is extracted, the extraction filtered and evaporated by half. Then the evaporation bowl is placed in the cellar. Crystals and various colors will sprout, upon which pour the above-mentioned spirit, and they will dissolve in it.

This Menstruum opens everything and resolves it into its prime matter, yes, in it gold melts like ice in hot water, leaving a white earth behind, and the solution looks like blood. Now pour this solution off into a clean glass, by inclination, seal the opening and preserve it carefully. The left-over white calx tastes like chalk and is of no use, except that it would be suitable if one wished to add a metallic earth to a mercurial work, because this calx or earth is quite simple.

The right Menstruum is drawn over in a retort to the consistency of a red oil. This Menstruum is also a pure Virgin of inestimable worth, which is well to be remembered.

Now take one part of this red oil - but before make another oil of native Hungarian sulphur in the same manner as the above-mentioned. Pour two parts of this latter oil to the first red oil in a Philosophical Egg in a B.M. Let it go through the colors *per gradus*, which will be unspeakably wonderful to behold. When they are coagulated together, the tincture can afterwards be augmented by the oil of the Sun and the oil of Mercury, and you will obtain a powerful tincture for Mercury.

But if this liquor or Menstruum is poured on a finely powdered golden ore in a retort - this ore can be found in *Chemnitz* and is a red clay with golden *paillettes* - so that it stands by 3 fingers' breadth above it, then a blind alembic is luted over it and it is set in mild heat, the ore or clay will begin to grow, but above there will be a thick fat red oil. The latter can be removed with a glass spoon every 4 weeks. Then it is put in a Philosophical Egg and congealed by a lamp.

When the 4 weeks are over, remove again the thick red oil above the Menstruum and the ore, and pour it to the other oil in the Philosophical Egg. In this way the Stone can be augmented *ad infinitum*, because every month a fairly good amount can be taken off.

In so doing, the following must however be heeded: Every 3 months some of the Menstruum, together with its crystals, must be poured into the flask which contains the ore. Thus you will obtain an eternal mine which can be used to augment the Stone in the Philosophical Egg. But some of the coagulated Stone must always be taken out of the Philosophical Egg before it is again imbibed with the red oil. Thus this Stone will finally tinge all metals into the very finest gold.

With this Menstruum one can also make big pearls, which is done as Follows:

Dissolve in this Menstruum a true live Mercury which will turn into a thick fat milk. Dip into it an oriental pearl which you would like to make bigger. Hang it in a small glass and put that in mild heat. Repeat this work after 24 hours and continue doing it till the pearl is big enough. Then dip it into the simple Menstruum whose sublimate is not yet dissolved, hang it once more into the glass, set it again in mild heat, and the pearl will get a wonderful lustre and, be more beautiful than the oriental ones.

This Menstruum, in which the live Mercury has been dissolved and which has turned into a milk, can also be coagulated by itself into a brilliant stone. The latter will then transmute the live Mercury into silver (because it has been prepared without oil of antimony). Or: If you wish to make pearls in a different way, prepare a fuming spirit in the following manner:

FUMING SPIRIT

Rx. 3 Lots of English tin, 5 Lots of live Mercury. Make an amalgam of them, pulverize and mix it with 8 Lots of a true sublimated Mercury. Pound them quickly together and immediately put the powder into a big retort with a wide but short neck (otherwise the powder will absorb the air), lute a big receptacle on to it, half of which must lie in water. Drive the spirit over by the grades of the fire, and it will come over very clear and with such might that you will be astonished. But if you see some salt that sublimates, stop. As soon as this spirit stands in the open and feels the air, it begins to fume strongly, and if the opening of the glass is not well closed, it will all go up in the air in smoke.

Take 2 parts of this first Menstruum, together with its crystals, pour it into a high flask and carefully add one part of the fuming spirit. You will be amazed how many colors will appear. This will last more than one month, until at last the substance turns into a milky liquid.

USE THAT AS FOLLOWS:

Dip a bristle into pure distilled dew water, so that one drop will cling to it. Gingerly insert this bristle with the drop into the flask, over this milky liquor. The drop will coagulate and change into a transparent pearl. Leave that attached to the bristle and take another bristle. Proceed with that as with the first transparent pearl - and make as many as you wish. After this, dip one pearl after another into this liquid, turn them well about, according to whether you wish to make them round or oblong. Then remove them again, dry them in mild heat, and dip them again into the Liquor. Continue doing this till they seem to be big enough to you.

NB. Before, however, you must have at hand a cross full of small holes, to allow you to insert into them the bristles with the pearls. Dip each pearl once more into the liquor, then stick them into the cross, set the cross into a broken-off wide flask, lute its opening very carefully- and set it, in mild heat, Thus they will ripen and become much more beautiful than oriental pearls. These pearls can now be given a color in the following way:

Just pour some oil of antimony to then, the processing of which will follow at the end, and they will obtain various colors.

But if some oil of the Sun is poured on them instead of the oil of antimony, they will become red like blood.

NB. This liquor coagulates all water into a milky stone.

With this liquor all precious stones can be imitated.

In addition, various works can be undertaken with this Universal Liquor before anything foreign is added to it and before it is used up, for instance:

A crocus of Venus (copper) is made as follows:

Take the flowers of verdigris of copper and distill a spirit out of them. Calcine the residual *caput mortum* (Death's Head) in an open fire (still better: First leach the salt from it.) Then dry the residue and pour enough milky liquor over this powder to stand 2 fingers' breadth above it. Thereafter put the flask, well luted, in a gentle digestion, and a thick oil will form above the liquor. With this oil all metals can be tinged into gold.

If this Regulus is poured on a Regulus *antimonii martialem*, (iron Regulus of antimony), the Regulus will instantaneously be dissolved. This solution is then carefully poured into another retort which is luted and set in a cool cellar, and crystals will sprout like the most beautiful diamonds.

Remove these crystals, take 2 parts of them, add to them one part of the above-mentioned thick oil, mix them well in a glass mortar, and they will give off a wonderful fragrance.

Now put the mixture into a phial, seal it hermetically, put the glass over a lamp, and it will go through the colors in one month. When it has become fixed, it must be augmented with this oil.

The Stone is an excellent medicine for men and metals, and it can be prepared with just this liquor.

With this milky liquor, prepared with its salt and without any other additive, (or with the subsequent liquor), in addition to the fuming spirit and the siliceous liquor, precious stones can be made.

How this liquor can also be obtained from vitriol pyrites, iron pebbles, auripigment, native cinnabar, haematite, talc, compact galena NB.

First: From all vitriol pyrites which have not been in any fire, the metallic spirit is made as follows:

Take this pyrite fresh out of the mines and immediately put it in casks to prevent it from attracting the astral, become heated, and thereby lose what is best.

If now you wish to melt this pyrite, fill half of a retort with it, set it in a sand cupel with a big receptacle luted on, give a mild fire to begin with, and a milky liquid will go over. After 2 days, increase the fire by another grade, and so on for 5 days. Finally, a strong acid goes over and much sublimate adheres to the retort and the neck of the receptacle. After this, let the fire go down and the retort cool. Thereupon remove the receptacle and mix the sublimate well with the liquor or spirit. Now the liquor is prepared. But if you wish to rectify this spirit, pour it into a big flask, lute on it an alembic and a big receptacle, and set it in the B.M. Begin with a very gentle fire, and the metallic spirit will rise over. Preserve it very carefully in a strong glass with a ground stopper, coat the grooves with cement, then set it in a cold place. This spirit is suitable for precious stones.

However, if you wish to prepare this liquor from sulphur Pyrates:

Take fresh pyrites and half fill a retort with them, add a big receptacle, then begin to distill very gently. In the beginning a milky spirit will go over, followed by strong spirits; finally some oil and simultaneously much red sublimate. Continue with a strong fire to make everything red-hot. Thereupon, let the fire go out, remove the receptacle and pour the evil-smelling liquor into a flask, lute an alembic and a receptacle on, set the flask in a B.M., and at first give a mild fire. A spirit will go over and a snow-white sublimate will form on top. Remove the receptacle, put this spirit in a strong glass in a quiet place. Screw the glass tight with a ground stopper and lute the grooves. This spirit can also be used for precious stones.

If now you wish to have the red sulphur oil at the same time, which is used for the preceding process:

Remove the flask immediately from the B.M., together with the liquor, set it in sand and resume distilling from the beginning. An acid spirit will go over together with some oil. The real oil of sulphur, however, will be left. Preserve this oil carefully in the meantime. NB. After this, extract the salt from the *caput Mortuum* (Death's Head), purify it, pour it into a retort, pour the oil on it, distill it over once more, and repeat this distillation till all the salt has gone over.

Now pour this oil into a big phial, seal it hermetically, coagulate it by the grades of the lamp, and you will obtain a red garnet which will easily melt.

Thereafter remove it from the phial, put one part wrapped in wax on 3 parts of gold in flux, cover the crucible with a strong cover and heat strongly till the gold turns into a red powder.

Put this powder on 10 parts of *Luna Cornua* in flux, and it, will also turn into a fat red glass. Put one part of this, wrapped in wax, on 100 parts of fuming live Mercury, and it will be transformed into the finest gold in a quarter of an hour.

If you wish to prepare this spirit from auripigment, however, you have to remember that a blood-red sublimate results from this distillation. Keep this from getting into the oil. Instead, take it out of the neck of the retort and sublimate it several times by itself. Then add it to the spirit which went over in the B.M., and keep it to work with precious stones.

But if you wish to proceed further with it, work with it as with the sulphur pyrites.

If you wish to prepare this liquor from native cinnabar:

Fill half a retort with native cinnabar, lute a big receptacle on, distill a spirit in the preceding manner, and finally much sublimate will rise over. When the spirit has all gone over, let the fire go out, and when the retort has got cold, remove the receptacle, add the sublimate to the spirit, mix them well together, then pour the spirit in a retort, distill it over again together with the sublimate, and you will obtain a golden-yellow smoke and spirit.

When you have 2 lbs., of it, take half a pound of Mercury purified and distilled over with quicklime. Put the live Mercury in a flask, pour upon it one lb. of spirit, lute the opening and set it in an ash cupel for 4 weeks. When you see that the Mercury is completely dissolved, put the retort in a sand cupel, and a big receptacle and begin to distill *per gradus* till everything has gone over in the form of a golden-yellow fiery spirit, and a white-greyish earth is left at the bottom. Discard the latter.

Pour this spirit again on one lb. of fresh live Mercury and proceed with it as with the preceding spirit. You will obtain a fat heavy Menstruum, and it is suitable for precious stones. NB. NB.

With this Menstruum one can thereafter reduce all metals, in particular gold, into their prime matter, which is done as follows:

Prepare a gold or silver calx - the purer the better - especially if it is made with Mercury. Put this calx in a flask, pour upon it twice as much Menstruum as the weight of the calx, lute the retort, set it in a gentle digestion, and the gold will melt like ice in hot water, again leaving a white earth at the bottom.

Then open the flask, carefully pour the extraction in a retort, discard the white earth, lute a receptacle on, put the retort in sand and distill the Menstruum back over. You can again use this Menstruum as before.

Finally, give strong fire, and your gold-sulphur will flow together into a transparent red glass. Then let the fire go out, break the retort open, put it in equal parts on gold in flux, and it will change the latter into red glass.

Of this, put one part on 16 parts of *Luna cornua* in flux, which will then turn into a red powder. Of this red powder put one part on 100 parts of live Mercury. When it is warm, it will all change into pure gold in a quarter of an hour.

With silver, it is done as follows:

This liquor is the true *Alkahest*, and if it should lose its power through frequent use, pour it again on one lb. of fresh live Mercury and distill it over. Then it is again as precious as before.

If you now wish to ripen unripe precious stones - which is a great secret - proceed as follows:

Take one part of the milky liquor, 2 parts of the smoking spirit, and 2 parts of the fat siliceous liquor. Pour all three together very carefully by drops, and they will coagulate together like a milk of many colors. This work must be done in a flask.

If then you have precious stones that are unripe, put them in this mixed liquor, lute the opening, and set the flask in mild heat for one month. They will become ripe and get the right color. They will also grow in it and become even bigger.

In this earth all things will grow, if you put a crystal in it, leaving it there for some time, it will form a skin like an egg, in which it will grow and become hard like the very finest diamond.

But if you add one part of the gold-oil of the first process, it will finally result in a tinging Stone, capable of transmuting silver into gold.

This then is what I wished to reveal to you, my Brother, about the transmutation of all things in the mineral realm. Work hard, call on God the All-highest, and the Mother of God, for wisdom, blessings and help, and you will perform miracles. But do not forget your poor neighbor, and use it to the honor of God. Amen.

NOW FOLLOWS THE PREPARATION OF THE SILICEOUS LIQUOR

See to it that you obtain white transparent pebbles, but not too big ones. Such are carried in large amounts in the Saale, the Elster, the Elbe, and other rivers. Pound them into a fine powder, mix among each pound 4 Lots of *Sal Ammoniack*, put in every retort one lb. and 4 Lots of *Sal Ammoniack*, lute a receptacle on, and melt them *per gradus* till the retort melts. The longer the retorts can stand it, the more spirit you will obtain and the fatter it will be.

When the retorts are all melting, let the fire go out. Then you will obtain a wonderful sublimate in the neck of the retorts, a golden-yellow one in front, a blood-red one in the center, and a grey one in the back. Something special lies hidden behind this sublimate. Put it into the grassgreen Liquor, pour the latter into a high flask and draw its water off to a fat oil in the B.M. Many consider this something special, in particular if this liquor is

208

added to the spirit of wine, when it will become a great medicine for stones and podagra[12]. And with this you have also got the right liquor *silicum* (siliceous liquor) for precious stones.

But if you wish to coagulate this liquor without spirit of wine into a salt, it will be for you a field into which you can sow whatever you wish, for everything will grow in it and become fixed.

If you add to this liquor the right grey sulphur, mix it together and coagulate it, then it is the right salamander which cannot be consumed by fire.

This salamander tinges lead and silver into gold.

THE GREY SULPHUR, HOWEVER, IS PREPARED AS FOLLOWS:

Rx. Antimony, arsenic, sulphur, one lb., equal parts. Pound everything small, mix it together, put it in a retort, and a blood-like arsenic will sprout on top. Remove this from the neck of the retort, add to it its weight in fresh antimony and sulphur, mix the powders again, put them in another retort and

---

[12] The photocopy from Hans is unclear here. A letter (or two) is missing after "pod" and before "agra". -pnw

give strong fire for 2 hours. Then let the fire go out.

Break the retort in two and separate the red arsenic. Pulverize it and add to it one part of pulverized, antimony and sulphur. Again put the powder into a retort etc.

You must repeat this sublimation, as I have here prescribed it, in all 7 times. With each sublimation you must add equal parts of antimony and sulphur. After the last one give fire, thereafter break the retort, separate the blood-red sulphur, NB., what is left under this sublimate is a Regulus - preserve it.

Pound the red sulphur small, put it in a large pan and calcine it till everything combustible has vanished, leaving a greyish-white powder.

This then is the grey sulphur which must be united with the pyrite salt. But if you united a crocus of iron or copper with this fat salt in equal weight and congeal them together for some time, and afterwards put this congealed substance in a Luna cornua in flux, the silver will yield much gold after the separation.

Powder Bohemian garnets, pour this siliceous liquor upon them, digest them, then again pour some more siliceous liquor on them and once more digest them - after that put the substance on lead, melt everything together, finally refine it on the cupel, and much gold will result.

The Following is also Pertinent.

Take one lb. of common Aquafort or spirit of niter, made with glue. Pour it over one lb. of vitriol calcined to whiteness, put it in a retort, and distill the Aquafort off by slow grades, so that the vitriol is not calcined, etc.

NB. The rest is indicated textually in the printed *Aurea Catena Homeri*, (The Golden Chain of Homer), Part II, page 355.[13]

NB. It dissolves all acid and alkaline subjects. For the white *astris*, one takes Aquafort or spirit of vitriol, distills it over alum, just as the above-described vitriol.

PROCESS

If now you wish to make of this a Salt-Stone, proceed as follows:

Take some ore from the mountain, whichever you wish, pound it fine and small, wash and pulp it, detonate it as usually, be it hard or easy to melt. Then let it become red-hot in a crucible, spread

---

[13] The *Golden Chain of Homer* is available as Vol. 33 of the R.A.M.S. Library of Alchemy. It is highly recommended. -pnw

some pulverized sulphur over it and mix it together
with an iron wire till the sulphur has been burnt
out. Now take one part of this prepared ore, put it
in a flask, pour one-quarter of the above-mentioned.
Menstruum of vitriol on it, if it is on white, pour
3 parts of the Menstruum distilled over alum. Set it
in the sand to digest and dissolve whatever can be
dissolved. Pour off gently and neatly from what is
not dissolved. After this, pour again 3 times its
weight of Menstruum on it, let it dissolve again and
digest, and continue doing this till everything has
become a clear liquor. Distill this to one-third in
a retort or a flask, let it cool, set it in the
cellar and let it sprout into crystals. Distill qnd
crystallize the Phlegma again to one-third, and you
will have a vitriol and *materia prima illius minerae
remotae*.

Dissolve this vitriol again into 3 parts of
Menstruum, distill it through the retort and
cohobate till everything has gone over. Now you have
a primordial vaporous liquor which cannot be
processed further without being harmed. And here you
have the mineral with all its Principles, because it
has never been robbed of its sulphur or marcasite,
much less of its metal, but all its vital spirits
and life forces have been turned into a liquor, for
if, e.g., you take refined gold, silver, tin, iron,
copper, etc., you have but one part, as in this

process of melting and refining it has lost its vital spirit and nutrient, which is the acid of vitriol, the sulphur, arsenic and marcasite.

If now you wish to distill, coagulate and congeal this liquor or oil, take a flask that is not too low, put it into it and digest it by boiling it in the A.M., till nothing rises any longer. Then set it in ash and distill all the phlegma per First, Second, and the Third gradus. Take the residue out, put it in a phial, set it in ash to coagulate, and it will turn into a salty Stone which flows in the fire like oil and, stands up in the air like ice. You must not stopper the phial, as the substance does not rise easily. And thus you have the Quintessence of the mineral, but it is still corrosive and cannot be used in the human body.

When you have made your minerals and metals (which have been reduced to their prime matter) volatile again, you must again make them fixed, according to the saying of the Philosophers: *Fax fixum volatile, et volatile fixum*. (Make the fixed volatile, and the volatile fixed), and you have the whole Art.

Therefore, take one part of your volatilized and again coagulated stone and add to it 12 parts of a compact body, be it gold, silver, copper, tin, etc. Lock it in a phial, set it in a mild digestion till the salt is united with the Body and has become fixed. Then add some more, digest and congeal it further, and continue doing this till the Salt Stone has turned everything into its own nature and has likewise changed the Body into a salty Stone which flows like wax in heat and stands like ice in the cold. Of this, take one part to 10 parts of metal in flux, and you will obtain a tincture. Of that, put again one part on 10 parts of metal, and proceed as I have described, and you will lack neither gold nor silver. Use it for the honor of God and the benefit of your neighbour, do not become arrogant and do not misuse it either but thank the Great God for His gifts.

NB. Instead of a compact Body it is still better to take some coagulated Mercury that has been dissolved with the vinegar containing the *Luna cornua* or the hell stone. Thereafter pour this solution over the live Mercury, which will finally swell up and be coagulated with it. In this way you will obtain a far-reaching silver tincture, as the subtle lunar parts are kept soft and porous by the Mercury, and the aforesaid metallic liquor conjoins and congeals with it the easier and more intimately.

NB. In this Menstruum I have dissolved, digested and, as taught above, distilled Mercury till a yellowish-brown oil is left at the bottom. I clarified this oil in the sand, rectifying it, twice per gradus with good spirit of wine. Finally, I distilled all the Phlegma off in a B. M., and thus I obtained the right oil of Mercury or spirit of Mercury as a beautiful crystal left at the bottom.

I put one lb. of it in a Pelican, added 4 Lots of Mercury, sealed it, and thus it also dissolved the Mercury into oil. In this way it, can be infinitely multiplied. The oil not only dissolves gold and silver but it also makes all extracted tinctures volatile and carries them over along with it. I took solar sulphur, poured on it 3 times the weight of this oil, let it putrefy for 14 days, distilled it through the retort, and as not everything went over, I again poured the same amount of oil over it, putrefied and distilled it as before, and I repeated this till everything had gone over.

The following Processes, which are prepared with butyrum of antimony and gold and silver, are to be considered for the polychrest.

ALCHEMICAL USE

Rx. One lb. of equal parts of Hungarian antimony and sublimated Mercury. Distlll from them an oil through the retort by sideways distillation.

These aforementioned items must previously be moistened with distilled vinegar. When the oil has gone over, take 12 Lots of extremely finely laminated gold or solar calx, precipitated with a solution of vitriol. Put it in a round glass and pour upon it all the aforementioned distilled oil, lute the glass carefully, then set it in sieved ash and let it stand in mild heat till the gold has been completely dissolved. Now it has become spiritual. After this let the glass stand in the heat till it is coagulated and has turned into a red-brown fixed Stone.

NB. In the *Secrets of Antimony* by Lemery, page 153, one can read that common antimony has always resulted in more butyrum than mineral antimony or antimony ore (idem p. 143), and that a mixture of 3 parts of common antimony and 4 parts of sublimated Mercury is to be preferred over all others.

Now open the phial, weigh its substance, then grate it small and pour 3 times the weight of the above-mentioned oil of antimony over it, seal the glass hermetically and proceed in the prescribed manner, coagulating and dissolving five more times. Now remove the coagulated substance, which is again as hard as a stone, pound it small, pour borax water over it - one little finger's breadth - coagulate it again, and the powder is now ready for the tincture.

NB. Conversely, concerning the preparation of cinnabar on page 167, Lemery has noted that more beautiful and better cinnabar of antimony can be sublimated NB. from antimony ore, that this cinnabar attaches to the neck of the retort in much larger and thicker pieces and that, according to the proportion, the mixture of 3 parts of antimony and 4 parts of sublimated Mercury generally results in more cinnabar.

Now take 20 Lots of fine silver, let it melt and gradually add one Lot of this fixed tinging powder, wrapped in wax and divided into 4 parts. Let both flow together for 3 hours, and you have the best gold. But, for the sake of curiosity, take some of this tinging powder before any borax water has been added to it, put some of it in a small flask, set it in gentle heat, and it will soon melt into a blood-red oil. Take a little of that - only the size of the head of a pin - with a golden spoon or a slate-pencil, put it on a silver plate the thickness of a Thaler (an old German coin). Put the latter in a crucible, wrong side down, so that the place where it is coated with red oil is underneath, but in such a way that it lies hollow in the crucible. Light a fire beneath to make everything red-hot, but without melting. The silver is thus penetrated through and through and will turn into the finest gold, as far as the oil has run.

Additional Observations

The oil that was first distilled must afterwards be rectified 3 times more before it is mixed with the gold.

The borax water is made as follows:

Distill the water of the pulverized Venetian borax through a retort, or dissolve the borax and calx in hot water, filter the solution, boil it to dryness, then let it flow to an oil in a humid place by deliquifying it. Or: dissolve the borax in some spirit of wine and then imbibe with it several times the mixture to be melted.

Among this and similar processes must also be counted that told me by an old Doctor and very experienced chymist, D. Petermann who assured me that a certain augmentation in gold and silver could be obtained with butyrum of antimony. To accomplish this, it would be necessary to take well rectified butyrum of antimony, prepared from a Regulus of iron or copper by means of sublimated Mercury abstracted several times from gold, then poured over silver calx and digested several months, the longer the better, and finally coagulated. This would certainly result in an increase in gold after its reduction.

And Mr. Schwanenberg, a very experienced chymist, has likewise told and assured me that, based on his own experience, he has worked out the known *Bezoardicum Lunare*, a silver counter-poison, according to the prescription of D. Agricolae, which he teaches in his commentary of *Poppium*, page 150; which deals with the chymical medicines prepared from silver. He also assured me that he had found a considerable increase or augmentation in silver after its reduction.

D. Kellner, in his treatise on lead processes, page 116, describes clearly and in detail *Posten's* tin opus, which Becher in his *Chemical Rosegarden*, p. 801, and others elsewhere recommend as a sure work.

Note. For the use of physicians, I would recommend that when a butyrum of antimony is precipitated with spirit of wine, the Mercury that has been dropped makes a subtle emetic which causes sleep and perspiration after the operation. See Stahl, in the German *Chymia Rationali*, page 433.

And precisely there, page 802, No. 18, a special work is described regarding the preceding foundation, which D. Stahl and after him D. Juncker, recommends- in his *Chymistry* above all others. It is to be done as follows:

Rx. Gold 2 Lots, resolve it in *aqua regia*; silver 6 Lots, dissolve it in Aquafort. Pour both solutions together and one will precipitate the other, and if they do not precipitate each other enough, add to them some spirit of salt or simply a solution of common salt. When everything is sufficiently precipitated, bring them both together to boil, then let them stand undisturbed day and night to allow the precipitate to settle. It will meanwhile grow tike black-berries. Filter it, edulcorate the remaining calx, dry it and ripen it by means of its inner active factor to a tincture, in the following manner:

The dissolved metals weighed 8 Lots. The increase from the saline spirits will be 1/2 Lot. Therefore add to this mixture 4 Lots of well pulverized Regulus of antimony, mix it and distill it through a retort per gradus. You will obtain a solar and lunar butyrum of antimony, or *arietem mineralem*. This butyrum must be rectified, which will turn it into a bright-red oil, suitable for ripening gold & silver, and iron lamellae or calx into a tincture in their whole substance, because this butyrum contains the Souls of the Sun, the Moon and Antimony.

Instead of taking the spirit of salt or salt, a good effect will be had if the precipitation is done with the salt of the calx of wine or with the salt from the Death's Head which is left after the sublimation of Mercury sublimated according to my prescription.

Take the goldish oil and digest in it the laminae of the Sun and the Moon contained in it. They will dissolve and coagulate into a Stone. The saline parts, however, which do not belong to the metallic composition, will gradually separate from it, which saline atoms were called "nymph ore" by the Philosophers. It would be very useful and helpful to the Work if that separation could be easily achieved, which is indeed possible and, as I believe, can be done with distilled vinegar or spirit of wine.

PROCESS FOR TURNING ONE MARK OF SILVER (GERMAN MONEY) INTO 4 LOTS *OF GOLD!*

A process of an experienced chymist, which is in accord previous one, is the following:

First, prepare the sulphur of iron with the help of the ammoniac, then precipitate, edulcorate, and dry it. After that, take equal parts of quite pure and crystalline sublimated Mercury and best Hungarian antimony ore (or pure antimony). Pound both small and mix them together, then melt them per gradus in the sand, and an oil will appear, at first whitish-yellow and finally, with a fire strong enough to make the sand red-hot, blood-red. Of this oil take 10 parts, of the above-mentioned sulphur of iron 3 parts, of a subtle calx of gold one part. Pour the latter two together in a small flask and pour upon it the 10 parts of the oil. Close the opening of the glass very tight, then set it in the steam bath for 3 days and nights, then in ash, and finally in sand, till it is successively coagulated. Thus it will turn into a brown Stone which pulverize, moisten with borax water and dry again. Repeat this 3 times and it will quite easily combine with molten silver. Afterwards, when it has thus been flowing for several hours, you will find 4 Lots of the best gold in the separation of one Mark of the thus transformed silver. Following this, the left-over and reduced silver can again readily be used. I have myself experienced this in action.

NB. If you pour some spirit of tartar, distilled from one lb. of crude tartar and 2 lbs. of vitriol over a crocus of Mars and Venus (iron and copper), letting it digest for several days and nights, that crocus will have a good ingress into silver and give off gold. Likewise: If this spirit is poured over an iron Regulus of antimony, it will soon extract a beautiful tincture. If the latter's spirit is extracted to the consistency of oil, a dark red liquor is left behind which, when poured over silver calx, soon turns black, and after a few weeks of digestion, the silver is quite goldish if it is melted and separated.

NB. If the gold is previously made liquid, it will supposedly resolve sooner and better in the butyrum of antimony and have a better effect. How to prepare such a ready liquidity is well described in D. Becher's "Chymical Glucks-Hafen", pg. 267, as may also be learned in D. Kellner 's "Officina Metall", page 73, ff .

Finis

MIXTURA PRAECIPUA MAGISTRALIS

Concerning the Divine Polychrestuous Salt Whose Inventor is the American Prince:

Haldazor A. Dachem
A Philosopher Without Equal!

Translated from the German tract:

"Geheimnisse Einiger Philosophen und Adepten, aus der Verlassenschaft eines Alten Mannes"
edited by: C. Hilscher, 1780

by: Léone Muller

Rx. Purest gold, 1 dram; purest silver, 2 drams; best bismuth ore, 1 oz.

First, melt the gold and the silver together, then gradually add some pulverized bismuth ore, let them flow together like water. Then remove the crucible from the fire and let the matter cool, and you will obtain a Regulus.

Pound this Regulus to a fine powder, mix it with one oz. of pulverized oriental pearls, put the powder in a retort, pour upon it some philosophically rectified spirit of vitriol q.s., set the glass in a mild digestion till the spirit has completely dissolved the powder. Then draw it over strongly, and a calx of various colors will be left. Remove this calx from the glass and put it in a flask, pour upon it enough distilled May-dew water to dissolve the calx completely. After that, filter the solution and precipitate it with a rectified spirit of vitriol. A white salt will fall to the bottom of the glass (or you can also just draw the spirit over to the consistency of salt). You must dissolve this salt several times in May dew, then evaporate it till it leaves no more feces after the dissolution. Then you have prepared it.

The virtues of this salt for the human body are almost incomprehensible. If, that is, 3 grains of this salt are put in a filter and you add one oz. of distilled plantain water or most highly rectified spirit of wine, the water or the spirit will be streaked blood-red. This imbibing with fresh water or spirit is to be repeated 6 times till the salt is totally opened. Then you have proceeded in the right way. Of this water every day one oz. is given to a mortally-ill person, and the most dangerous illness is radically cured after the seventh dose, when no other medicament could help.

NB. Of the blood-red splrit, however, 10 to 15 drops are given the Patient in e spoonful of wine, and he will be completely restored after the seventh dose. If one drop of this spirit is poured one red-hot silver plate, it will tinge it into the most beautiful gold.

Virtues vero hujus Salis, ad Metalla imperfecta perficienda, mirificae sunt. The virtues of this salt for the perfection of imperfect metals are miraculous.

After precipitation with spirit of vitriol and its crystallization, this salt must not be distilled, dissolved, and again crystallized like the first water of May dew, but it is processed in such a way as to make it magnetic for attracting the astral sunbeams and making them fixed.

No. 1. Fill some strong flat evaporation bowls, spreading this salt by half a finger's breadth in them. Then wait for the hottest days of the summer to put them in the sun in the morning, so that the latter may attract the Phlegma from the salt. Thereafter, keep this salt in a strong glass with a ground stopper in a cold place.

No. 2. When April or May arrives, take this salt from the glass, put it in an evaporation bowl, set that at night in the air to allow the moon to shine on it, and this salt will magnetically attract some water from the air, which will melt it. In this water is hidden the general vital food of the air.

No. 3. The following morning, set the bowl with the water, uncovered, in the rays of the sun, letting the water evaporate again. In this way, it will retain only the goldish seed of the sun and let go of the powerless water.

No. 4. The bowl must be set in the sun - but by no means in the moon light at night - till it has attracted all the water. The Philosophers call this dissolution and coagulation a philosophical day.

No. 5. The dissolution and coagulation of this salt must be repeated several hundred times, till it no longer attracts any moisture from the moon at night but stays dry; and if some of it is put on a capper plate to test it, it must melt on it without fumes and tinge the copper into the very finest gold as far as it has spread out. Then one has worked properly.

No. 6. If a man is so lucky as to bring this salt to perfection from the month of April to the end of August, the merciful God has been gracious and merciful to this man than to many thousand others.

No. 7. Finally, take this salt out of the evaporation bowl while it is dry, put it in a strong phial, seal the latter hermetically, and set the glass in the rays of the sun from the end of August, and in the light of the moon at night (but take care that no rain touches the glass) till all the colours appear in it and at last the salt becomes blood-red. Then one has obtained the greatest treasure on earth.

No. 8. If you wish to transmute silver into gold with this salt, proceed as follows:

Dissolve 1000 parts of fine silver in Aquafort, precipitate it with a good spirit of salt, and a white calx will result. Pour a large quantity of warm water on it till the calx has settled. Afterwards, pour the water off and add an equal amount of warm water. Continue doing this till the water finally tastes quite sweet. Then the calx has been faultlessly prepared for the subsequent work.

No. 9. Put this silver calx in an evaporation bowl, and in the center one part of the red salt. Put the glass uncovered in the light of the moon at night and in the rays of the sun during the day, and due to the celestial magnet the silver calx will dissolve into a water in one philosophical day. Proceed with this water as I have taught in the first process, and this silver calx will all become a tincture.

No. 10. Now divide this tincture into two equal parts. With one part tin and live Mercury are tinged into stable gold, in the following manner:

Melt 1000 parts of tin, lead, or live Mercury in a crucible, or heat the Mercury. Then put one part of the tincture, wrapped in wax, on the flowing tin, lead, or fuming Mercury, let it flow well covered for a quarter of an hour before the bellows, then pour it into a hot coated ingot, and the metal will be transmuted into the finest gold.

No. 11. With the other part of the tincture you can ferment fresh silver calx infinitely, making a tincture of it, if you mix the tincture with 1000 parts of fresh silver calx, proceeding further as has already been taught. **Laudetur Deus T. O. M. in secula seculorum, Amen.** (Praise be to God for ever and ever, Amen.)

Eques ab Oriente, A.C.R.C. Illustrissiums. Leipzig, 1736.

HIS MAGNETIC SALT

Rx. Melt 1/2 oz. of the very finest silver, and while it is in flux add 2 oz. of pulverized bismuth ore. When the two substances are well combined (but beware of their fumes, because they are arsenical), let the crucible cool, pound the Regulus you have obtained into fine powder, put it in a flask, pour upon it some strong spirit of niter mixed with 3 parts of distilled vinegar, lute a blind alembic on the opening, set the glass in a mild digestion for a few days stir the glass frequently, then pour the Menstruum off by inclination. Again add fresh Menstruum and continue doing this till the Menstruum no longer tastes vitriolic. Pour the extractions you have obtained together upon the Death's Head, add some distilled water of May dew q.s., boil it well, then pour the boiled and filtered water to the extractions you have kept and evaporate them *ad cuticulam*. Set the evaporation bowl in a cold place, dry the crystals you have obtained, preserve them in a clean glass with a ground stopper in a cold place and you have prepared the magnetic salt.

EXPERIMENT

Take one gr. of salt, put it in a big sugar glass that is filled with 2 cans of river water, set the glass in the open between 11 and 12 noon, in May, June, July, or August, to allow the sunbeams to concentrate in it. After 2 hours you will see with amazement how the water gradually turns blood-red, though as soon as a cloud passes before the sun, the redness sinks to the center of the water, covering the surface with a skin. The water will look white above and below, but after sundown a brown earth will precipitate to the bottom of the glass. Let the water stand over it for another day, then pour it off by inclination, dry the earth, mix it with wax and put it on silver, refine it with lead, separate it in Aquafort, and you will obtain a fine grain of gold which has taken its origin from the sunbeams.

HIS UNIVERSAL SALT

Take whole pieces of Spiegel Russ (mirror Soot, Specula), put them in a bowl in the open air for some time, then put the Spiegel Russ in a pot, connect the latter with a bladder into which poke holes with a strong needle, and bury the pot in the earth. Let it stand there for several months, and you will find some white salt on the Spiegel Russ. Collect this carefully in a clean glass, screw it tight with a ground stopper and set it in a cold place. With this salt you will achieve great cures. Regarding this, read the 15th. Introduction to the curious **Bucher - und Staats-Cabinet**, year 1719.

HIS TINCTURE FOR PARTICULARS

With it he tinged copper into the very finest gold in the presence of the blessed late King of Poland, Frederick August I., who had been at that time sick in bed in Leipzig.

The tincture consisted of the following species:

Process.

Take 10 parts of the best Mercury, put the powder into a small glass flask, do not lute the opening, set it on coal in a cooking pot filled with sand, and give fire gradually till the Mercury begins to flow. Keep it in this condition from 6 to 8 hours and white flowers will sublimate in the neck of the flask. In the center, however, there will be a metallic sublimate which looks like sewing needles and is quite heavy and compact. Keep this sublimate provisionally carefully in a clean glass.

THE METHOD OF PREPARING A RED OIL FROM PHOSPHOROUS AND GOLD IS NOW GIVEN AS FOLLOWS:

Put one oz. of phosphorus in a wide-necked flask, pour upon it one dram of gold calx opened in the way taught later. Lute the opening, set the glass in a mild fire, and the phosphorus will dissolve the gold. Now open the glass, set it in the open air at night, and the phosphorus will have liquefied and radically opened the gold, transforming it into a red oil.

Now put 10 parts of the flowers of the live Mercury into a small glass, imbibe them with the red oil till you have put all of it on the flowers (you would do better to mix this red oil with the flowers in a glass mortar and then to put the substance in the flask), lute an alembic with a small receptacle on the opening but it must not be luted tight, so that the spirit cannot push it off. Set the flask in a sand cupel, give fire per gradus, and red glassy flowers will be sublimated, which will not ignite and will disappear with the smoke when projected on silver or other metal.

Put one part of these red flowers on a red-hot copper plate. Put the latter on a coal fire and the flowers will spread on it like oil deeply penetrating the copper, and they will tinge it into the very finest gold.

NB. The gold calx for this process is made as follows:

Make an amalgam of half an ounce of the finest gold and 3 oz. of live Mercury. Put this in a flask, pour upon it common Aquafort q.s., and it will dissolve the Mercury, leaving the gold as a yellow mass. Edulcorate and dry this.

If one dram of these red flowers is put on 10 parts of *Luna cornua* in flux, it will enter them like oil. Put this silver on 10 parts of silver in flux, refine and separate it, and you will get 5 drams of the finest gold from one dram of this flower. **Sapienti satis.**

THE MYSTERY OF URINE

HOW TO PREPARE A TINCTURE FROM IT BY WHICH OTHER METALS CAN BE TRANSFORMED INTO GOLD!

First, the red salt of urine is prepared as follows:

Let good urine stand for some time till it putrefies, then distill from it the spirit of urine by itself, while evaporating the remaining Phlegma to dryness.

Pour this spirit of urine on this salt residue in a flask, cover it with a blind alembic and set it for several weeks in a mild digestion. Then the spirit of urine extracts and dissolves the salt and sulphur contained in the residue. This done, collect the spirit of urine in another clean flask upon which apply an alembic with a receptacle luted on it. Then distill the spirit of urine gently till a salty skin appears. That seen, set it in the cold, and it will result in beautiful bright red, yes, dark red transparent crystals. Carefully pound these with gold leaves, and finally melt this mass in a crucible. The gold therein will be completely opened and turned into a glassy substance, almost that of a ruby. If this substance is melted with silver and separated, it is supposed to tinge many parts of gold. But if it is extracted, it is supposed to be an incomparably good medicine.

HOW TO MAKE THE PHILOSOPHER'S STONE FROM THESE RED CRYSTALS

Take one part of sulphur of the Sun - which has already been described - mix it with one part of the red crystals, put this powder in a phial, seal it hermetically, and set it in the B.M., for 91 days. Then it will be congealed and flow like wax without fumes, and one part will tinge 13 parts of Mercury into gold.

Pound this mass again to a fine powder and mix it with equal parts of the red crystals, seal the glass hermetically, set it in the B.M. for 91 days, and the tincture will congeal, and one part will tinge 25 parts of Mercury into gold. Continue as before with the addition of red crystals till the tincture congeals in 8 days, and finally in 3 days. Then one part of it will tinge two hundred thousand parts into good gold.

PHOSPHOROUS TAKEN FROM HUMAN URINE

Take 15 parts of a good spirit of salt, pour into it one part of spirit of niter, and the *aqua regis* (royal water) is ready.

Rx. Now take fine gold, a half ounce, and 12 oz. of this Aqua Regis. When the gold is completely dissolved in it, abstract the Aqua Regis so much that it takes on a thick and oily consistency. Then add half a dram of good phosphorus and gradually abstract this oily mixture through a spacious retort per gradus. A big receptacle must be luted on the retort and must have a small opening, as it would otherwise be thrown off.

When the phosphorus has combined with it, you can lute the opening tight, and all the gold will go over as a fine distilling butyrum. If it should not have risen over completely the first time, pour some fresh Menstruum and phosphorus on the residual gold and abstract again, and a fine butyrum will go over completely. You must rectify it twice more by itself and keep it well preserved.

Now prepare the flowers of live Mercury as follows:

Put enough live Mercury in a flask with a wide and long neck to leave 3 parts of it empty. Set the flask in a sand cupel and do not close the opening. Gradually give fire till the live Mercury begins to flow. Keep it in this flux for 10 hours, and white shining flowers will sublimate, followed by a metallic sublimate which looks like sewing needles and can easily be melted.

Of these flowers, better sublimate, take 10 parts and pour on them 4 parts of the butyrum which you have kept. Thereupon seal the phial hermetically and set it in the B.M., till the substance has gone from the blackness to the whiteness.

After this, remove the phial from the B.M., set it in a previously heated ash cupel, give the 2nd., 3rd., and 4th. grade of fire till the substance has turned red like blood. Then let the fire go out and the glass cool down.

Now open the glass, take the red substance out, weigh it, pulverize it, and put it again in a new and strong phial. Pour over it twice the weight of the butyrum you have kept, and seal the opening hermetically. Proceed as has already been described, and this stone will multiply from 10 to 10^{14} times, provided you weigh it, pound, and congeal it each time as prescribed, N.B., with the butyrum of the Sun.

But if you wish to ferment it with gold and make of it a Stone for other metals, proceed as follows:

Rx. Melt 1 1/2 oz. of gold and put some of this red powder, wrapped in wax, on half an ounce of the gold in flux. Let them flow together in a well-

---

[14] Perhaps the original said "10 to 100" or some other numbers, but the R.A.M.S. edition clearly says "10 to 10". -pnw

covered Hessian[15] crucible for 6 hours, and the gold will turn into red glass.

Retain half of this red glass for augmentation. With the other half tinge as follows:

Take half an ounce of this red glass, pulverize it, mix it with wax and put it on 1000 Lots of lead, tin, or copper in flux, and let them flow together in a strong fire for 3 hours, and you will obtain the very finest gold.

Pulverize the other half of the Stone which you have kept, put the powder in a phial, pour upon it 2 parts of the butyrum of the sun, seal it hermetically and proceed as has been taught in the process. In this way you can multiply the Stone infinitely.

Rx. Put 4 oz. of English phosphorus in a big sugar glass, cover it with a linen cloth and set it in the open air at night; during the day, however, in a humid cellar till the phosphorus is completely dissolved and has turned into clear water which will weigh twice as much.

In a glass flask pour this water over 4 oz. of live Mercury with which you must previously have mixed 1/2 oz. of sulphur of the Sun. First allow the water to permeate the powder, then lute an alembic

---

[15] A Hessian crucible is a type of ceramic crucible that was manufactured in the Hesse region of Germany from the late Middle Ages through the Renaissance period. They were renowned for their ability to withstand very high temperatures, rapid changes in temperature, and strong reagents.

with a receptacle on the glass - but lute only the alembic, not the receptacle - set the flask in a sand cupel, give fire per gradus, and bright red flowers will sublimate. These are glassy and do not ignite in the air. Collect them carefully in a clean glass and keep them. When you wish to tinge with them, proceed as follows:

Heat a copper plate red-hot, put one part of these flowers on it and they will immediately melt, penetrate the copper and tinge it into the very finest gold as far as they have permeated.

Process.

Dissolve the gold in spirit of salt mixed with the 15th. part of spirit of niter. In it dissolve 1/2 oz. of gold and abstract the Menstruum till it turns oily. Into this oily substance put half a dram of good phosphorus and about a pinch of coal dust, lute a big receptacle on the retort - but there must be a small opening as otherwise the receptacle will be thrown off. When the phosphorus has combined with the other substances, the opening can be luted again. Thereafter, gradually distill the Menstruum off to dryness, and the gold will all be sublimated, or rather distilled, into a butyrum. If everything has not gone over the first time, pour on some more of the spirit that gone over, again add some of the phosphorus, and it will all go over as a butyrum. When this butyrum is dissolved in fresh distilled rainwater, it releases a white earth. The rest must not be reduced. After this, the dissolved butyrum must be filtered, gently evaporated, and the residue then sublimated. In this way the gold or any other metal is cleansed of its useless and superfluous earth and its color is heightened. If it is afterwards mixed with a purified Mercury and congealed per gradus, it is supposed to be suitable for tinging.

In this way all metals can be made volatile and distilled over. Then silver and live Mercury that have been dissolved in Aquafort and treated with phosphorus result in a beautiful white and shining butyrum, and the silver will finally sublimate quite white, like pearls.

This volatilization of metals is considered very arcane, as with such a volatilized gold, worked into a butyrum and mixed with equal parts of a thus treated live Mercury and then congealed together, many parts of other metals in flux have been tinged. See Juncker's *Chemistry*, page 882.

A PHILOSOPHICAL WORK WITH PHOSPHOROUS

Take 3 to 6 oz. of the best phosphorus, put it in a big glass pot covered with a white linen cloth, set it at night in the full moonlight, but remove it before sunrise and keep it in a dark place. This operation is repeated several times till all the sulphur has been completely dissolved into viscous water through common Mercury and salty mercurial water, and has got almost twice its previous weight.

Pour this watery fire and fiery water in a cylindrical white glass. Carefully lute on its rim and opening a specially ground burning glass with a small hole in the center, about the size of a pinhead. Add an ox bladder. Then take a big flat concave mirror, about 15 to 20 inches in diameter, set it firmly on a stand, so that it can always be turned according to the course of the sun. Put this mirror against the first glass to allow the focus to fall through the small burning-glass into the vessel and the radical moisture of the whole nature - not in its greatest force though close to it.

Here you can see to your greatest amazement how this greedy magnet often attracts the solar atoms in their loveliest colours and absorbs them corporeally. Yes, this red masculine sulphur, like strong rain and dew, invades its lunary-magnetic *Enixum* so often that the glass, too, is sometimes completely filled with it, at which, in strong sunlight, you become aware how an ignition occurs in the glass at certain times, like lightning from heaven. The result is a horrible vapor and a strong emanating smell, and the focus must be arranged somewhat weaker when this sign appears.

Continue this operation till it begins to take on a reddish color. N.B. The glass in which the substance is kept must be covered with a slightly moistened cloth or else stand in the shade.

In the meantime, the *Corpus Solis* (Body of the Sun) is also prepared by this magical fire in the following way:

Take gold purified to the highest degree q.v. Amalgamate it with Virginian or purified Mercury, let it stand in digestion for 24 hours, then distill the Mercury from the gold with a big concave mirror which must rest on a stand. The Mercury will disappear in smoke.

Now make another amalgam with the gold powder and the Virginian Mercury. Put it in a phial to digest for 24 hours and again calcine the gold with the rays of the sun. Continue this work till the gold powder has turned from red to dark-purple, which is completed in 2 months. Finally, this gold powder is given a saline form.

At the point I took my above-mentioned prepared reddish Menstruum, gradually poured it on the gold powder through a phial and let it stand in gentle digestion for some time. After putting all the Menstruum on it by and by, always leaving the phial in mild digestion for one day, the mass swelled up considerably. Thereupon I increased the fire for 3 days and nights till the Menstruum was completely dried up. Finally, I melted 3 parts of the very finest silver, put this powder on it, wrapped in wax, and let them flow together. Then everything turned into the very finest gold. Thus did I work it.

If I had had a larger supply of this reddish Menstruum, I could later have turned it into a tinging oil and improved the latter's quality, which would have further resulted in a multiplication of the quantity.

END OF THE FIRST PART

THE UNIVERSAL PROCESS

(Being a Treatise on Dew)
A dying Cappucine monk left this tract to
His beloved brother and signed it with his
Blood. Prague 3-29-1672

In the Name of God, take some of the whitest
sea salt which the ships bring from the Island of
St. Huber in Spain. First it should be calcined by
the rays of the sun. Let this salt dry in a warm
room, then grind it in a glass mortar to a fine
powder and dissolve it in dew.

During the month of May or June, when there is
a full moon, observe when dew is gathering while the
wind is blowing from the east or south during the
day or the night before. Go to a field and drive
some stakes of wood into the ground about 1 1/2 feet
deep and in the shape of a triangle. This way, you
can put a sheet of glass on top of the stakes. The
dew can gather on this in the morning. Under it,
have a few glass containers handy into which the dew
can flow. You have to repeat this with many
containers in order to collect enough dew. The full
moon quarter is best for this, afterwards it does
not have enough strength. Put the gathered dew into
glass containers, lute them and store them in the
cellar until you need them.

When you want to start this holy process, put as much as you want of the retained dew into a flask. Little by little, add some of the sea-salt powder. Continue with this until the water becomes saturated and you see undissolved sea-salt at the bottom of the flask. This is the way it should be because the dew will have the right weight.

Take a 1/2 pound of this solution and pour it into a short-necked phial, filling it half way. Close the opening with a glass stopper and lute it so that the universal spirit cannot escape. Put the phial into a water bath so that the dew can putrify. The dew has to digest for 42 days and nights (a "Philosophical Month") over a low flame. You will see the matter blacken which is the sign of Putrefaction. (Crow's Head) put the phial into a *Balneum Siccum* and coagulate the water over a gentle flame. This will take place within 14 days and nights. You will see a gray salt coating the sides and bottom of the vessel. Before this becomes too dry, remove it from the flame and let the phial cool a little. Then replace the phial back into the water bath over a gentle flame for 40 days and nights after which the salt will dissolve again into the water. When it blackens again, put it back into the *Balneum Siccum* so the water can coagulate again. As soon as you see the grey salt appearing, let it cool and put back in the water bath. Repeat this process for a total of five times. The water should be light and clear as crystal and, after coagulation, the salt will be as white as snow.

Put some of it onto a glowing metal plate. When it flows like wax it is tested (proven). Do not take the salt out of the phial; put it back into the water bath. It will dissolve again into water but will be reduced by one third. Instead of salt water, you will have a sweet and drinkable water, the greatest medicine on earth. Give 25 drops of this to a person and he will be cured of the most serious of illnesses.

There are much greater secrets in this water, but I am sworn to silence. You will be able, yourself, to recognize enough of this power of Heaven and earth. *Sapiente Satis*.

If you want to make a tincture of metals, proceed as follows: pour as much water into the phial as you need, put it well-luted in an ash-cupel over such a gentle flame so that the glass gets only warm. Open it and add little by little, some gold oxide or some finely laminated gold leaf, until the water cannot absorb any more. This you will be able to observe the following day, (you can proceed the same way with silver, getting a white tincture - the work is the same). The water now has the right weight with the gold added. Pour this solution, without the undissolved gold (decant) into a clean

phial, about 1/4 full. Seal it hermetically, put it
into a water bath and let it stand over a gentle
flame 40 days and nights. You will observe a lot of
black matter. As soon as this is observed, put it
into the Balneum Siccum, once again placing it over
a gentle flame until you hear a noise like water and
ice and see some beautiful colours and, most
remarkably also, the creation of the world! After
the 12th. or 13th. day, it will change into a
reddish-brown powder tike red Cinnabar. The White
Tincture, however, will transmute into a crystalline
matter.

This red powder you can project onto metals in
the following manner: Take five (5) parts of gold or
silver, encase one part of the powder in wax and put
it over the metal in a sealed container. This you
stand over high heat for one hour. When you take the
container off the fire, the gold will be brittle.
Encase one part of this brittle gold in wax of
inferiour metal in a liquid state. After one hour,
this will have transformed into best gold.
I advise you to be careful with this powder, do not
waste it. If you put this reddish-brown powder in a
phial in the water bath, you will have a red oil
within 35 days and nights. With silver you will have
a light blue oil. If you take 3 drops of this oil
with champagne, it can heal all wounds, it radically
cures all illnesses, it will keep a human healthy to

the end of his days by making him lose all his hair and nails and then making them grow back again, new and youthful. This also will cure everything by creating a high temperature and lost strength will be regained within a short time. *Quod per Deum testar possum.*

If you put one end of a silver coin into the red oil, it will transform this part into gold immediately without harming the print. If you want to project the oil still further, put it again into the Balneum Siccum. Within 10 days it will transform into a powder with the most beautiful colours, only, much redder and prettier than before, glowing like a ruby or carbuncle with silver, the powder will look like snow.

If you put one part of this onto 50 parts of Molten Gold (if using silver powder, use molten silver) and let this flow together vigorously for one hour. The Gold or silver will transform into a tincture. Of this, encase one part in wax and add it to 100 parts of molten metal over a high flame. You will have fine gold after one hour.

If you want to further augment your tincture, put the powder in a phial into a water bath for a 3rd. time. Within 30 days you will have a dark red oil from gold and a white oil from silver.

One single drop of this in wine will perform the already described wonders. You only take the dose two times a year, as this medicine is very fiery. You have to be careful because both body and soul are affected.

Continue with this oil a fourth time like before, putting the phial into the Balneum Siccum. The oil will coagulate again. You will observe all kinds of colours and living things moving up and down. In the end it will become a dark red powder again.

Encase one part of this powder in wax and add it to 500 parts of molten gold (or, if using the white tincture, to molten silver). The gold or silver will transform into a tincture. Again, encase one part of this tincture in wax and add it to 1000 parts of inferior molten metal, after one hour you will have the finest of gold or silver.

You will have to continue five times with this dissolution and coagulation of the first medicine. After the fourth coagulation you will have a tincture. Of this take one part and add to it 5000 parts of molten gold (or silver) transforming it into a tincture. Take one part of the tincture and encase it in wax and add it to 10,000 parts inferior metal in molten state. After an hour, it will all be the finest gold (or silver),

You take the first powder or tincture, if, without fermenting it further and, after dissolving and coagulating it five times, you strengthen and augment it, within 24 hours you will have an incredibly red powder or, from the white, a white crystalline powder. Of this last coagulated powder put one part with a molten fifty mark gold coin (or white on silver). After one hour, the gold or silver will be a tincture. Put one part of this with 100,000 parts of molten inferiour metal. After one hour it will be exquisite gold.

I got so far with this and no further. Otherwise it will seep through the glass and disappear into the air with the most delicious smell.

Take note, during your work, many Fratres *Rosea Crucis* will come to you because you caused them to see it.

Observe, I wrote down this holy secret and swear by it on by belief of the Holy Ghost and Jesus Christ. I sign this with my blood on my death-bed, on my last day on earth.

Finis.

THE SPAGYRIC ART

Johan Tritemius, Abbott of Sponheim

**Unity, he says, is not a number
But all numbers arise from it.
Translated from Theatrum Chemicum by Pat Tahil**

Before the universal water of the abyss that is
mentioned in Genesis was divided, it was one. By
this division alone, the one produced two, the first
of all the numbers, not in essential substance, but
by the chance of circumstance. It is a number and is
quantified, yet it is not a number and is not
quantified. It is not quantified because it is
single in character, and it is quantified in so far
as it is composed of chance happenings. However, it
cannot so far be quantified because there is no
number previous to it. His understanding of unity is
that which Hermes declared in other words, and he
explains it by a simile taken from Genesis. For it
is supernal, he says, and the rest follow. For two
is defined by Hermes as above and below. Trithemius
says, If two is thrown away, then three will be
converted to one. Hermes says the same in other
words, for carrying out the miracles of the one
thing. Here it must be particularly noted that unity
is distinguished from either of two in two ways,
because Hermes counted unity as above and below, as
we have said, while Trithemius defined the first

number as one. Later, for very different reasons, either of two defines unity, one by the reduction of above and below, like a miracle, two by the rejection of two and the (subsequent) conversion of three to one. It is truly wonderful that both agree. There is therefore a certain natural unity, divisible into, of rather, enumerable as, two, and three can be turned into the other unity, which is called the second unity, beyond which it is not proper to proceed.

As all the operations of nature within its limits consist of wonders, it descended from unity through the double to the triple, not, however, before the quadruple had arisen by a simple series of steps. For if you wish to count to four, you can only begin with one, and you say One, Two, Three, Four, which taken together make ten. This is the perfect consummation of all number, for then there is a return to one, and there is no simple number above ten. The ignorant will wonder at the profundity of this connection, which we use as the principle responsible for performing miracles, or whether we have the help of demons, or whether we are superstitiously relying on conjectures contrary to our Christian faith. We, however, judging these things for ourselves on account of the ignorance of those who preceded us, do not wonder. For as Holy Writ tells us that no one except him who has received it can understand the internal experience of God, so whether one is versed in these things, or whether one cannot make use of it unless by divine favor he has received a unique insight into understanding nature by nature, there will be fire in him besides the light, a wind with the fire, power with the wind, and with the power, knowledge and purity of mind; for this is the foundation of deep matters, and the root of all creation. The first division of nature produced the root of sound science, to which I make this note. There are four mothers for those in the latest order, and four fathers for those in the first, the binding of all these, and the first logical connection, the final elemental pure one-and-only, alone pervading all things. Earth is an element pure and simple, the first to proceed from the one, it is not compounded, it is not changed, it does not suffer combination, but remains as it is, incorruptible, and one consists of one, yet not one, it is not a number, yet it is a number, it is not quantified, and it is quantified, between one and itself there is no number. One remains unity, and by union it makes three; eight times by including this reduces all to one, by a marvel of nature; no teacher is able to explain its power over all.

It is not the same as the God whom we worship; it is an image created in the mind of man, neither alive nor dead that produces wonderful effects for all kinds of knowledge. And I tell you nothing but God's truth my friend, for whoever is uplifted by the sheer idea of this pure simplicity will be perfect in all natural science, will perform wonderful works, and will discover remarkable results. The single one-and-only is good, and from it flow not only like things, but many dissimilar things. Compound earth is a natural element, pure, simple, and unique, but because it is compound it is necessarily varied, multiple, and impure, but can nevertheless can be reduced by fire to water, from this to fire and from this to homogeneous unity, and it is a number and is counted, and it is not a number and is not quantified. It is not quantified because it is of an uncompounded nature, and compound only by chance occurrence. Therefore it cannot be quantified because there is no number before it. A one that is not absolute but inclusive is counted after unity, and one says One exclusive, One inclusive, and one by means of one from the one, (that is, the soul of the world) and three is arrived at: this naturally wishes to be with the one: one powerful in itself, but impotent either as one or the other, always rolls into a ball; one remains as a fire, but, not such as one can conceive of. If it is reduced to its elemental state by purification with fire and suitable washing, it can perform all the mysteries of profound knowledge.

The compound earth is an element and not an element, through which two is reduced to three and four, distinct from one by several steps. It contains wonders, it is manifold, and multiple, and corruptible, yet it does not stray outside the circle of unity: the mastery of all secrets is this, as well as (knowing how) three becomes one by the agency of two. And whatever wonderful human inventions exist, they are subject to its power, and can be effected by the complete operation. It respects number, degree and order, by means of which nature performs all its miracles. It is able to perform marvels, and more things than one can believe that require neither insult to God, nor a spotted soul. Through it wonderful works are done: through it is obtained full knowledge of all human inventions, and effective performance in secret matters, for its power, proceeding from understanding, does not allow the worker to go astray for three stages. It discerns everything men say. A task begun with it cannot be led astray into error. Whatever astronomers, mathematicians, magicians, alchemists, those jealous investigators of nature, do, and whatever the worse kind of necromancers promise demons, it can discern, rectify, understand, discover, and, prepared by its beginnings, bring to completion without wickedness; it is an element, yet not an element, removed from itself by number, joined to itself by simply reducing to unity. Without knowing by numbers stages and degrees its middle, beginning and end, no magician can give power to his imaginings without wickedness, nor can he do so even if he performs unrighteousness: no alchemist can imitate nature, no man can bind spirits, nor can prophets foretell the future of the world, nor can any inquisitive person understand the reason for his experiments.

And so the workings of all nature existing within its limits consist of miracles; it descends from unity to the tertiary by way of the binary, not however, before it has risen from the quaternary to oneness by a series of steps, as was said above, etc. Natural knowledge, therefore, which sometimes consisted of pure simplicity, founded upon natural principles, was mixed with so many lies and impurities and so much deception, that there is no one unless he is extremely learned in both natural and supernal science who can distinguish the one from the other, or understand it. I myself have known (he says) so many men go astray in natural philosophy, however learned, of whom some by devoting themselves to alchemy lost time and money, some lose their life together with their goods, others, seeking to make a medicine by it, produce nothing after long labors, others, seeking wonderfully powerful secrets, can arrive at no result, others, eagerly desiring to foretell the future, speak lies instead of truth, others, carry empty notions in their heads, having recently read records that they do not understand; they ignorantly produce the good and the true, and the bad and the false. Accordingly, these three principles of this natural, Spagyric and occult philosophy, without perfect knowledge of which no worker can produce results, must first be known and declared.

The very beginning consists of one, through which, not from which, the power of the miracles of nature produces its effect, of which we said, "because the purity that comes from unity does not form a compound, nor does it change". There is a progression to it, the monad, from the ternary and the quaternary, in order to complete (the number) ten: from that there is a regression to the number one, as well as a descent to four, and an ascent to the monad. Ten can only be completed by itself: the one is gladly converted to the three. All those who do not know this beginning after the beginning with unity neither accomplish anything in the ternary, nor reach the sacred quaternary. For even if they have all the philosophers' books, and know the paths of the stars, their powers, their abilities, and their workings perfectly, and although they may understand their images, rings, signs and great secrets to the full, nevertheless they can produce no miraculous effects in their works without knowing this beginning from the beginning in the beginning. Therefore however many practitioners of natural philosophy there are, either they have achieved nothing, or have fallen into vanity, frivolity and superstition in desperation after long and useless endeavors. Indeed, this second beginning, separated from the first by degree, but not by superiority, because the coming into existence of one makes three, is that which works wonders by means of the binary. For there is unity and non-unity in one, it is uncompounded, yet compounded of four; when it is purified by fire into gold, pure water comes forth, and when it has returned to its pure state it will show the worker the accomplishment of secrets. This is the centre of natural wisdom, whose circumference is joined to itself; its vast arrangement recalls the circle to an infinite degree: its power when purified is above all, but when elementary is less than anything compounded beyond the fourth degree.

However, the Pythagorean number four supported by three, if it observes order and degree, when purified to absolute unity from the triple in the twofold can perform wonderful secrets of nature. This is the tetrad within which the two in the three combined with one make a whole that performs miracles. For three is a number reduced to one by the power of sight, it contains all in iteslf and can do what it likes.

The third beginning is not a beginning in itself, but between it and two is the boundary of all science and mystic art, and the undisputed center of the middle: it is the easiest place to make a mistake, for there are very few living in the world who understand its depths. It is variable and composite, and by means of seven it becomes eight times three and remains fixed. In itself it is the perfection of number, order, and degree, and by its means all philosophers and true investigators of the secrets of nature achieve wonderful results, by it reduced to the simple element in three ways wonderful cures of ill health and natural sicknesses are effected, and the work of the practitioner of natural and supernatural knowledge produces its results. Demons flee from the regular arrangement of four. Prediction of the future is verified by it, and in the nature of things secrets are penetrated by it alone. By this means alone is the secret of nature laid bare to the alchemists, without it no knowledge of the art is acquired, nor does the work reach any result. Believe me, he says, they err, they all err if they think they can do any work on the secrets of natural science without these three beginnings. It is however, a great source of error that the sages of yore who were possessed of the secrets of nature either said nothing about them or hid them so very obscurely that no one except their peers could understand them. The secret and heavenly philosophy of this arrangement is that if anyone really wishes to know and understand it, he must flee human turmoil, put the world aside, and contemplate the heavens not only with his eyes, but also with his mind; the spirit of God bloweth where it listeth and lightens whom it will, and whomsoever it shadows with its power it leads to full knowledge of the truth. Let him who has received it give thanks to God, and be eager to repay the gifts he has received with the fruits of good works, and let him realize he has received them, and not find there any reason for pride. However, let him to whom it has not been given to know these sublimities either realize the weakness of his intellect, for his striving did not bring him knowledge, or praise the compassion of his creator who created obstacles to his knowing for his own good: and if he has not acquired this knowledge, let him say that he is not in debt to God for such a gift, and not complain. Thou art my friend, he says, to hear thy friend taking thought for thee, and thou knowest what thou hast done. Victor over thy neighbors tread down the fire of envy, not mortal aspirations, which is an insult to immortal God: levity brings danger. Flee the gatherings of men and worldly cares, bind thyself to heavenly meditation. Thy safety is in the swallow's nest, thy peril in hens' dung. Thou shalt follow a veil borne away on the wind, seven times shalt thou be wearied, but thou shalt rise again to unity by means of the three, and find thyself fortunate. If thou hast begun thy work with Sol, which in nature appears to set, so that thou shalt turn from all to God, the true Sun, with zeal for knowledge, thy mind purified from lower things, with desire in thy soul, in the fervor of sacred love, and he shall show thee another. For the art of divine love is long lasting, while time is short, and it is better for the creator to cherish truth rather than his creatures. These are Trithemius' words, showing others that he supported the Hermetic art.

It is true, as Hermes says, certain without a word of a lie, and most certain by acquaintance with unity. What is below is like that which is above, and the reverse, seeing that every number is made up of single ones, for performing the many miracles of the one substance. Does not everything flow from the single goodness of One, and whatever is united to one cannot be of a different nature, but gives fruit in the simpleness of and ty the adaptation of one. What is obtained from one, except three? Listen, One is simple, two is composite, three can be reduced to

the simpleness of one. I am not Trithemius the
triple-minded, but of one mind taking delight in the
number three, and that indeed gives birth to a
wonderful child. His father is the Sun, his mother
the Moon. The air carried the seed in its womb, the
earth nourished it. This is the father of all that
is perfect in the world. His power would be
uncorrupted and vast if indeed he existed on the
earth. Thou shalt separate the earth from the fire,
the thin from the thick and three, already gone back
into itself with great skill and gentleness, shall
rise from earth to heaven, and then, adorned with
power and beauty, shall return to earth and shall
receive strength from above and below and shall be
powerful and glorious in the brightness of unity,
ready to produce all numbers, and all darkness shall
flee away. Three must therefore be reduced totally
to one, if one's mind desires to reach a full
understanding of these matters. For unity is not a
number, but all numbers arise from it. As one
withdraws from one, two is the first composite
number. Two may therefore be rejected and three
will be converted to the simpleness of one. All
numbers consist of single units. Does not everything
flow from the goodness of one, see above, and
whatever is joined to one cannot be of a different
nature but brings forth fruit in the simpleness of
and by the adaptation of one? What is born of one,

except three? Unity therefore is simple, two composite, three reduces to the simpleness of one. Unity is pure beginning. Leaving one behind, two is composite, because it is impossible for there to be two beginnings. Therefore three alone is sacred, powerful, and virtuous, and two, having been excelled, returns to its beginning, not naturally, but by affinity: in this the mind sees no contradiction, and understands very well all the mysteries of a series of secrets. This is the beautiful virtue of courage, that conquers all worldly things. It is very true that certain things are necessary for any man who wishes to do useful work in the art, using natural wisdom.

Firstly it is necessary to have the right disposition for the art, not merely be inclined to it, or at least to have a master of the teachings at one's beck, because of the rectification of three into one by division by two. Secondly one ought to have sufficient command of language to prevent the common herd from understanding the great majesty of this science of natural wisdom. A knowledge of the fundamentals of astronomy is necessary, or at least one should have at hand someone who knows them. Thirdly, many books on this science are necessary, only those that have been most carefully corrected, such as are seldom found to-day, or someone at hand or ready to correct their mistakes, otherwise there is no profit. Fourthly a teacher learned and expert in this art is necessary, for the science is so wrapped up in mysteries that without a very skilled teacher one cannot understand, unless Almighty God wishes to illuminate the mind with extraordinary giftedness, which very seldom happens.

Fifthly, a knowledge of the division of the whole universe, superior and inferior from one into four, then its settling into three, is required. Likewise one ought to know the order of ascent and descent, degree, number, bending back and forth, existence and non-existence as one and as three; it is indeed very difficult to know this, for the whole origin of miraculous effects, by whatever means they are accomplished, whether by natural or supernatural wisdom, depends on this principle as a foundation; therefore all who understand this order and the means to acquire it, will be supreme in every science, and the depths of wisdom, and will achieve marvelous results for their labors. But since it is very difficult to know these things, there are very few who do useful work in the spagyric science and many who labor fruitlessly. Sixthly, one ought to learn a suitable mode of living, the order of the work, the hour of the day, the matter, and the ruler of the material, that is, its planet, the right place, form and material, and the mixing of materials, pure or impure, simple or composite, and how to bind those things that are conjoined: and after all this [one should learn] the capacity of one's mind, its strength and its inherent power of goodness. Seventhly, one ought to know under the rulership of which planet, spirit of the hour and spirit of the season exists with its substance, casual properties, and effect that mundane substance through which they operate best. For things below are subject to things above and are assimilated to one by resembling each other: this one exists in its substance, inessential qualities, power, strength, number, degree, and properties because of the application of the one to the other, and when it has been established by the art useful work may be done by the miracles of natural wisdom.

Eighthly, it is necessary for the worker in this craft to know and understand all the properties of the intelligences, their degree, place or location, names or words, and their function or work, how they exist at the ends of the series, how they may be used for working on any universal purpose. Otherwise one must first know how they are perfected for certain, as meat is preserved with salt so that it does not putrefy. Ninthly, one must have companions for what one cannot achieve alone, whether they be naturally worthy, or made worthy by their office, for one's comrades' unworthiness impairs work in any operation, whether of natural or supernatural wisdom.

Tenthly, the worker ought to have a firm belief in this art, and consequently have no doubt or hesitation about its results, not because belief in any way helps to attain these, but because doubt discourages the resolve of the worker to go from the middle to the end, and weakens him, and unless he is steadfast he will not have the desired inspiration from on high. Eleventhly, anyone wishing to work profitably with natural wisdom ought to keep everything highly secret, and reveal to none either the assistance of other men, or the failings of the worker, or the work itself, or its meaning, or the art, or the right time, except to a teacher or disciple: for this subject flees disclosure, and if divulged rarely yields perfect fruit.

This our philosophy is heavenly, not terrestrial, like that highest principle that we name God, so that by the mind's insight with faith and knowledge we may see the Father, Son and Holy Spirit, one principle, one God, and believe truly in the one greatest good existing forever in the persons of the Trinity, know, and forever adore with reverend service and the most fervent love Him from whom come all things which can exist anywhere. Unless the inspired mind rises to this [level] it will understand nothing that is excellent, but will waste away in its ignorance. This ascent is not for the common person, nor is imitation of those who are carried up on one wing at most sufficient, but [it is for him who] is familiar with the few, namely those who with him return neither wrongly nor rashly to unity. Many try, but not all have the triple mind. When we look at the sky must we not first raise our heads, and draw then back after we have looked up? It is given to the eyes alone to see the Sun, the ears do not see. Therefore, as the eye and the heart cause the soul to rise, not the ear, so unity is made by the triple participation of goodness in the beginning, for One is the all-powerful good, not Two or more. For unless One is made, no joining into its likeness can be made in the mind, nor can goodness take part, and there is no transcendence without these; for unless these come first, no one will be able to understand either the meaning of those things that are above, or how properly to employ those things that are below. Things universal as well as things particular are necessary, and [various] states of things, some of which are clear, some clearer, some extremely clear, and there are others that are obscure, more obscure, or most obscure, both to sense and to reason: in such diversity does nature operate. So it is that certain wiser men climb beyond the others. However, it is said of the wise man rather that he sees less of appearances. We are opening a way for the intelligence of these men. Accordingly, whoever aspires to learn either natural or supernatural philosophy can prepare his way into either of the two through the other, nevertheless, he will proceed more safely if he reduces himself from two to three by means of the fourth degree before he attempts or presumes to do so in the natural world. Although he can indeed arrive conversely at the supernatural through the medium of the alchemical medicine, this other way is more faulty, nor is there any in which error is more likely, therefore I would persuade everyone to choose to go the other way. Agree with this, excellent reader, and do not strain either your waiting soul or your ears in vain.

An Interpreter's clearer explanation.

Natural man is one and is not numbered, but Supernaturally he is counted as two, as spirit and body, which form a duality in him. Now on account of the original corruption the latter overcomes the other, so that the spirit cannot produce anything wonderful. For that to happen in this life, the double must be overcome by the triple, that is, the body must conform to the nature of spirit, and the spirit be joined to the body, so that it finds peace in it in turn. When this is done, the triple, already in existence, rejoices in the perfect secord unity; for by two previous commands one was united into three. Still, this conjunction should be made using the fourth degree, that is, by transmutation of the elements, of which the body is made, into a single highly purified element, in this way:-

First, a certain water is produced from the earth of your body, that is, your stony, earthy and sluggish heart becomes soft, and eager to know its God, and to reach Him: thus certain images and thoughts of the spirit can be impressed on it, like signs on wax. Afterwards, air is made from this water, that is, raise yourself upwards to heaven to Him who has created your humble and contrite heart. Like air that always tends to rise, and entreat Him with prayers to open your mind to the understanding of those things that come from God. Finally, fire is produced from this air, that is, your heart, already risen, turns its whole desire to love (to which fire is compared because of its heat) of God and your neighbour here on earth, so that its flame shall never be extinguished. In this day your two, namely spirit and body, are joined by affinity into a perfect three, using the three steps of the quadruple, as you have heard. Now, dear reader, you already have the key of the contemplative philosophy that opens the ascent to the heights, and none closes it against you. Of the way of descent that closes so that none can open it you shall hear in what follows. So you must again descend to the earth from the heaven to which you have ascended to receive the strengths of above and below. Behold how few are the things that alchemists both ancient and modern have tried to conceal under the wraps of so many and various riddles: I have tried to explain them to your free mind, and also to all students of things hidden by my meagre labors. Unwearied, I shall endeavor to expound to you many things that are closely connected with this philosophy, so long as you are watchful and attentive, which indeed is to your advantage. Hear more concerning the order of number and degree.

Number consists of order and measure. Nor can order exist without number and measure. Measure, however, consists of number and order. This unity-and-trinity does not allow of number, but, stripped of all plurality, consists of the first order in its own innate simple purity. This is the way to the gods, by which the ancient sages set out, guided by the light of understanding and reason, and learned much that is now considered beyond human knowledge by our own sages. Hear further: Study brings knowledge, knowledge gives birth to love, love to imitation, imitation to participation, participation to worth, worth to power, and power does miracles. This single way leads to the perfections of the adept's philosophy, both natural and supernatural, from which all that is superstitious, deceitful, and devilish is kept at a distance, and thrown into disorder. Since therefore, the goal of the contemplative science and science of the adept is truth, and the goal of the practical alchemist is the work, we know as much as we can understand of the knowledge of God, who alone is truth: we understand in proportion to our labor. For the true and healthful knowledge of God gives birth to understanding, understanding to love, love to companionship, companionship to trust, trust to the obtaining of all that you have asked for. Knowledge surely precedes the pursuit of virtue, for no one can long for what he does not know of. Recognition of truth and love of the right prepare the surest way to happiness. However, as we have said, recognition comes before love, for indeed no one can love what is unknown to him. Therefore our Lord Jesus Christ said of his father in the Gospel, This is eternal life, that they shall know thee to be the one God whom I have sent, Jesus Christ. For what is the height of the more-than-heavenly delight of the blessed spirits but the knowledge and love of the divine majesty? For a healthy understanding of science has love attached to it, nor can the perceptive mind have a part in the eternal benefits, if understanding is without love, and love without understanding.

Certain evil demons understand, but since they have no love they cannot reach the fruit that is born of both and not of either singly. Certain heathens outside Christianity today, perhaps even many of them, seen to have love of the highest truth, but since they do not know the one true God of all, and our lord Jesus Christ whom he sent, their thoughts are empty and they shall in no way attain the benefits of the highest good. Our Savior Jesus Christ himself said in the Holy Gospel of those who having come to him in this world do not recognize him. He who does not believe is already judged, for truly knowledge comes from faith, and love from knowledge. Therefore he who lacks it will never have knowledge. Moreover, he who has no knowledge lacks love, and he who does not love will be balked of reward. For this is what our Lord Jesus himself revealed to his disciples as he was about to ascend into Heaven, Go ye into all the world and preach the Gospel to every creature. He who believes and is baptized will be saved: he who does not believe will be condemned. So the way to God is first science or knowledge, through faith, without which no one shall be saved. All other sciences and studies must be referred to this true knowledge, because unless that is done, those wise men who distinguish their studies from those we have spoken of will be destroyed in vain and foolish labor. The true wisdom and knowledge that we have propounded greatly affect the knower with their goad, do not raise up [his spirits], do not allow him to be proud, but cause him to groan, according to the word of the wise man. He who adds to knowledge adds to sorrow, for in much learning there is much unworthiness. Therefore let us see that our studies are real ones, while we have the time. This is what he says. So that I can make clearer to you what I have explained to you concerning the ascent of the two to the triple by means of the four, observe:- The first step of the ascent upwards is striving for faith, for this disposes the heart of man to dissolve into water. The second step is the knowledge of God through faith, which disposes the contrite heart to ascend into the higher air, and the hope of a better life. The third step is love of God by knowledge through faith and hope, disposing the airy heart to love and charity and to the flaming fire of desire, through imitation of union with God. The fourth is constant company by which the love that has begun is continually refined through frequent meditation joined with prayer, cherished in growing faith, hope, and the love already conceived, and, the heart turning to the heavens is almost united with then. The fifth step is familiarity which is acquired ty constant dealings of this sort with the divine. The sixth is trust, by which we, made bold by the constant offering of faith, hope, and charity, dare to make our petitions personally to God, being certain that we shall obtain that which we desire from our Father. The seventh is getting those things we have asked for and desired in the name of Jesus Christ, by which we have the benefit of his glory, our salvation. Finally we are prepared for all mysteries, natural and supernatural, and as adepts of the philosophy created by God, are filled with all wisdom. It must certainly be noted that as our heart rises, leaving the vile, dirty and corruptible earth, in the first, second and third steps, so in the fourth, fifth, sixth and seventh it descends into a renewed, incorruptible, solid and constant earth, strong to resist the assaults of any enemy, and never again separable from the unity in which it is joined. Altering his words a little, Trithemius previously divided this ladder into nine steps, that can be interpreted, as above, into either the philosophy of the adept or the alchemist; for both exist as parts of one and the same wisdom. Theory of the higher or supernatural knowledge, practice of the lower natural wisdom. Since all knowledge of natural secrets depends upon knowledge of the supernatural, we shall interpret the said steps more easily and clearly by the contemplative philosophy, thus.

Tract on the Tincture and Oil of Antimony

by

Roger Bacon

(circa 1214 - 1292)

On the true and right Preparation of Stibium
to Heal Human Weaknesses and Illnesses therewith
and improve the imperfect metals.

From Friedrich Roth-Scholtz,

Deutsches theatrum chemicum,

Nürnberg: Adam Jonathan Felsecker, 1731.

Preface

Dear reader, at the end of his Tract on Vitriol, Roger Bacon mentions that because of the multiplication of the Tincture that is made from Vitriol, the lover of Art should acquaint himself with the *Tract De Oleo Stibii*. Therefore I considered that it would be good and useful that the *Tract De Oleo Stibii* follows next. And if one thoroughly ponders and compares these tinctures with one another, then I have no doubt that one will not finish without exceptional *profit*. Yet, every lover of Art, should mind always to keep one eye on Nature and the other on Art and manual labour. For, when these two do not stand together, then it is a lame work, as when someone thinks he can walk a long path on one leg only, which is easily seen to be impossible.

VALE

Joachim Tanckius

De Oleo Antimonii Tractatus.
ROGERII BACONIS ANGLI
Summi Philosophi & Chemici

Stibium, as the Philosophers say, is composed from the noble mineral Sulphur, and they have praised it as the black lead of the Wise. The Arabs in their language, have called it *Asinat vel Azinat*, the alchemists retain the name **Antimonium**. It will however lead to the consideration of high Secrets, if we seek and recognize the nature in which the Sun is exalted, as the Magi found that this mineral was attributed by God to the Constellation Aries, which is the first heavenly sign in which the Sun takes its exaltation or elevation to itself.

Although such things are thrown to the winds by common people, intelligent people ought to know and pay more attention to the fact that exactly at this point the infinitude of secrets may be partly contemplated with great profit and in part also explored. Many, but these are ignorant and unintelligent, are of the opinion that if they only had Stibium, they would get to it by Calcination, others by Sublimation, several by Reverberation and Extraction, and obtain its great Secret, Oil, and *Perfectum Medicinam*.

But I tell you, that here in this place nothing will help, whether Calcination, Sublimation, Reverberation nor Extraction, so that subsequently a perfect Extraction of metallic virtue that translates the inferior into the superior, may profitably come to pass or be accomplished. For such shall be impossible for you.

Do not let yourselves be confused by several of the philosophers who have written of such things, i.e., Geber, Albertus Magnus, Rhasis, Rupecilla, Aristoteles and many more of that kind. And this you should note. Yes, many say, that when one prepares Stibium to a glass, then the evil volatile Sulphur will be gone, and the Oil, which may be prepared from the glass, would be a very fixed oil, and would then truly give an ingress and Medicine of imperfect metals to perfection.

These words and opinions are perhaps good and right, but that it should be thus in fact and prove itself, this will not be. For I say to you truly, without any hidden speech; if you were to lose some of the above mentioned Sulphur by the preparation and the burning, as a small fire may easily damage it, so that you have lost the right penetrating spirit, which should make our whole Antimonii corpus into a perfect red oil, so that it also can ascend over the helm with a sweet smell and very beautiful colors and the whole body of this mineral with all its members, without loss of any weight, except for the foecum, shall be an oil and go over the helm.

And note also this: How would it be possible for the body to go into an oil, or give off its sweet oil, if it is put into the last essence and degree? For glass is in all things the outermost and least essence. For you shall know that all creatures at the end of the world, or on the last and coming judgement of the last day, shall become glass or a lovely amethyst and this according to the families of the twelve Patriarchs, as in the families of jewels which Hermes the Great describes in his book: As we have elaborately reported and taught in our book de Cabala.

You shall also know that you shall receive the perfect noble red oil, which serves for the translation of metals in vain, if you pour acetum correctum over the Antimonium and extract the redness.

Yes not even by Reverberation, and even if its manifold Beautiful colors show themselves, this will not make any difference and is not the right way. You may indeed obtain and make an oil out of it, but it has no perfect force and virtue for transmutation or translation of the imperfect metals into perfection itself. This you must certainly know.

AND NOW WE PROCEED TO THE MANUAL LABOR, AND THUS THE PRACTIC FOLLOWS.

Take in the Name of God and the Holy Trinity, fine and well cleansed Antimonii ore, which looks nice, white, pure and internally full of yellow rivulets or veins. It may also be full of red and blue colors and veins, which will be the best. Pound and grind to a fine powder and dissolve in a water or Aqua Regis, which will be described below, finely so that the water may conquer it.

And note that you should take it out quite soon after the solution so that the water may conquer it. And note that you should take it out quite soon after the solution so that the water will have no time to damage it, since it quickly dissolves the Antimonii Tincture. For in its nature our water is like the ostrich, which by its heat digests and consumes all iron; for given time, the water would consume it and burn it to naught, so that it would only remain as an idle yellow earth, and then it would be quite spoilt. Consider by comparison Luna, beautiful clean and pure, dissolved in this our water. And let it remain therein for no more than a single night when the water is still strong and full of Spirit. And I tell you, that your good Luna has then been fundamentally consumed and destroyed and brought to naught in this our water.

And if you want to reduce it to a pure corpus again, then you will not succeed, but it will remain for you as a pale yellow earth, and occasionally it may run together in the shape of a horn or white horseshoe, which may not be brought to a corpus by any art.

Therefore you must remember to take the Antimonium out as soon as possible after the Solution, and precipitate it and wash it after the custom of the alchemists, so that the matter with its perfect oil is not corroded and consumed by the water.

THE WATER; WHEREIN WE DISSOLVE THE ANTIMONIUM, IS MADE THUS:

Take Vitriol one and a half (alii 2. lb.) Sal armoniac one pound, Arinat (alii Alun) one half pound / Sal niter one and a half pound, Sal gemmae (alii Sal commune) one pound, Alumen crudum (alii Entali) one half pound. These are the species that belong to and should be taken for the Water to dissolve the Antimonium. Take these Species and mix them well among each other, and distill from this a water, at first rather slowly. For the Spiritus go with great force, more than in other strong waters. And beware of its spirits, for they are subtle and harmful in their penetration.[16]

When you now have the dissolved Antimony, clean and well sweetened, and its sharp waters washed out, so that you do not notice any sharpness any more, then put into a clean vial and overpour it with a good distilled vinegar. Then put the vial in Fimum Equinum, or Balneum Mariae, to putrefy forty (al.i four) days and nights, and it will dissolve and be extracted red as blood. Then take it out and examine how much remains to be dissolved, and decant the clear and pure, which will have a red colour, very cautiously into a glass flask. Then pour fresh

---

[16] Wear a face mask! -HWN

vinegar onto it, and put it into Digestion as
before, so that that which may have remained with
the faecibus, it should thus have ample time to
become dissolved. Then the faeces may be discarded,
for they are no longer useful, except for being
scattered over the earth and thrown away.

Afterwards pour all the solutions together into
a glass retort, put into Balneum Mariae, and distill
the sharp vinegar rather a fresh one, since the
former would be too weak, and the matter will very
quickly become dissolved by the vinegar. Distill it
off again, so that the matter remains quite dry.
Then take common distilled water and wash away all
sharpness, which has remained with the matter from
the vinegar, and then dry the matter in the sun, or
otherwise by a gentle fire, so that it becomes well
dried. It will then be fair to behold, and have a
bright red color.

The Philosophers, when they have thus prepared
our Antimonium in secret, have remarked how its
outermost nature and power has collapsed into its
interior, and its interior thrown out and has now
become an oil that lies hidden in its innermost and
depth, well prepared and ready. And henceforth it
cannot, unto the last judgement, be brought back to
its first essence. And this is true, for it has
become so subtle and volatile, that as soon as it

senses the power of fire, it flies away as a smoke with all its parts because of its volatility.

Several poor and common Laborers, when they have prepared the Antimonium thus, have taken one part out, to take care of their expenses, so that they may more easily do the rest of the work and complete it. They then mixed it with one part Salmiac, one part Vitro (alii. Nitro, alii. Titro), one part Rebohat, to cleanse the Corpera, and then proceeded to project this mixture onto a pure Lunam. And if the Luna was one Mark, they found two and a half Loth good gold after separation; sometimes even more. And therewith they had accomplished a work providing for their expenses, so that they might even better expect to attain to the Great Work. And the foolish called this a bringing into the Lunam, but they are mistaken. For such gold is not brought in by the Spiritibus (alii. Speciebus), but any Luna contains two Mark gold to the Loth, some even more. But this gold is united to the Lunar nature to such a degree that it may not be separated from it, neither by Aquafort, nor by common Antimonium, as the goldsmiths know. When however the just mentioned mixture is thrown onto the Lunam in flux, then such a separation takes place that the Luna quite readily gives away her implanted gold either in Aquafort or in Regal, and lets herself separate from it, strikes it to the ground and precipitates it, which would or might otherwise not happen. Therefore it is not a bringing into the Lunam, but a bringing out of the Luna.

But we are coming back to our Proposito and purpose of our work, for we wish to have the Oil, which has only been known and been acquainted with this magistry, and not by the foolish. When you then have the Antimonium well rubified according to the above given teaching, then you shall take a well rectified Spiritum vini, and pour it over the red powder of Antimony, put it in a gentle Balneum Mariae to dissolve for four days and nights, so that everything becomes well dissolved. If however something should remain behind, you overpour the same with fresh Spiritu vini, and put it into the Balneum Mariae again, as said before, and everything should become well dissolved. And in case there are some more faeces there, but there should be very little, do them away, for they are not useful for anything.

The Solutiones put into a glass retort, lute on a helm and connect it to a receiver, also well luted, to receive the Spiritus. Put it into Balneum Mariae. Thereafter you begin, in the Name of God, to distill very leisurely at a gentle heat, until all the Spiritus Vini has come over. You then pour the same Spiritum that you have drawn off, back onto the dry matter, and distill it over again as before.

And this pouring on and distilling off again, you continue so often until you see the Spiritum vini ascend and go over the helm in all kinds of colours. Then it is time to follow up with a strong fire, and a noble blood red Oleum will ascend, go through the tube of the helm and drip into the recipient. Truly, this is the most secret way of the Wise to distill the very highly praised oil of Antimonii, and it is a noble, powerful, fragrant oil of great virtue, as you will hear below in the following.

But here I wish to teach and instruct you who are poor and without means to expect the Great Work in another manner; not the way the ancients did it by separating the gold from the Luna. Therefore take this oil, one lot, [ancient weight unit used for the weighing of gold and silver coins - about 1/30 pound] eight lot of Saturn calcined according to art, and carefully imbibe the oil, drop by drop, while continuously stirring the calx Saturni. Then put it ten days and nights in the heat, in the furnace of secrets, and let the fire that this furnace contains, increase every other day by one degree. The first two days you give it the first degree of fire, the second two days you give it the second degree, and after four days and nights you put it into the third degree of fire and let it remain there for three days and nights. After these three days you open the window of the fourth degree, for which likewise three days and nights should be sufficient. Then take it out, and the top of the Saturnus becomes very beautiful and of a reddish yellow colour. This should be melted with Venetian Boreas. When this has been done, you will find that the power of our oil has changed it to good gold. Thus you will again have subsistence, so that you may better expect the Great Work.

We now come back to our purpose where we left it earlier. Above you have heard, and have been told to distill the Spiritum vini with the Oleum Antimonii over the helm into the recipient as well as the work of changing the Saturnum into gold. But now we wish to make haste and report about the second tinctural work. Here it will be necessary to separate the Spiritum vini from the oil again, and you shall know that it is done thus:

Take the mixture of oil and wine spirit put it into a retort, put on a helm, connect a receiver and place it all together into the Balneum Mariae. Then distill all the Spiritum vini from the oil, at a very gentle heat, until you are certain that no more Spiritus vini is to be found within this very precious oil. And this will be easy to check; for when you see several drops of Spiritu vini ascend over the helm and fall into the recipient, this is the sign that the Spiritus vini has become separated from the oil. Then remove the fire from the Balneo, though it was very small, so that it may cool all the sooner.

Now remove the recipient containing the Spiritu vini, and keep it in a safe place, for it is full of Spiritus which it has extracted from the oil and retained. It also contains admirable virtues, as you will hear hereafter.

But in the Balneo you will find the blessed bloodred Oleum Antimonii in the retort, which should be taken out very carefully. The helm must be very slowly removed, taking care to soften and wash off the Lute, so that no dirt falls down into the beautiful red oil and makes it turbid. This oil you must store with all possible precaution so that it receives no damage. For you now have a Heavenly Oil that shines on a dark night and emits light as from a glowing coal. And the reason for this is that its innermost power and soul has become thrown out unto the outermost, and the hidden soul is now revealed and shines through the pure body as a light through a lantern: Just as on Judgement Day our present invisible and internal souls will manifest through

291

our clarified bodies, that in this life are impure and dark, but the soul will then be revealed and seen unto the outermost of the body, and will shine as the bright sun.

Thus you now have two separate things: Both the Spirit of Wine full of force and wonder in the arts of the human body: And then the blessed red, noble, heavenly Oleum Antimonii, to translate all diseases of the imperfect metals to the Perfection of gold. And the power of the Spiritual Wine reaches very far and to great heights. For when it is rightly used according to the Art of Medicine: I tell you, you have a heavenly medicine to prevent and to cure all kinds of diseases and ailments of the human body. And its uses are thus, as follows:

AGAINST PODAGRA or GOUT

In the case of gout one should let three drops of this Spiritu vini, that has received the power of the Antimony, fall into a small glass of wine. This has to be taken by the patient on an empty stomach at the very moment in time when he sense the beginning or arrival of his trouble, bodily ailment and pain. On the next day and afterwards on the third day it should also be taken and used in the same way. On the first day it takes away all pain, however great it may be, and prevents swelling. On the second day it causes a sweat that is very inconstant, viscous and thick, that smells and tastes quite sour and offensive, and occurs mostly where the joints and limbs are attached. On the third day, regardless of whether any medicine has been taken, a purging takes place of the veins into the bowels, without any inconvenience, pain or grief. And this demonstrates a great power of Nature.

AGAINST LEPROSY

To begin with the patient is given six drops on an empty stomach. And arrange it so that the unclean person is alone without the company of any healthy people, in a separate and convenient place. For his whole body will soon begin to smoke and steam with a stinking mist or vapor. And on the second day his skin will start to flake and much uncleanliness will detach itself from his body. He should then have three more drops of the medicine ready, which he should take and use in solitude on the fourth day. Then on the eighth or ninth day, by means of this medicine and through the bestowal of Divine mercy and blessing, he will be completely cleansed and his health restored.

AGAINST APOPLEXIA OR STROKE

In the case of stroke, let a drop of the unadmixed tincture fall onto the tongue of the person in need. At once it will raise itself and distribute itself like a mist or smoke, and rectify and dissolve the struck part. But if the stroke has hit the body or other members, he should be given three drops at the same time in a glass of good wine, as previously taught in the case of Podagra.

AGAINST HYDROPE OR DROPSY

In the case of dropsy give one drop each day for six days in a row, in Aqua Melissae or Valerianae. On the seventh day give three drops in good wine. Then it is enough.

AGAINST EPILEPSIA, CATALEPSIA, & ANALEPSIA.

In case of the falling sickness, give him two drops at the beginning of the Paroxismi in Aqua Salviae, and after three hours again two drops. This will suffice. But if further symptoms should occur, then give him two more drops as above.

AGAINST HECTIC

In case of consumption and dehydration, give him two drops in Aqua Violarum the first day. On the second day, give him two more drops in good wine.

AGAINST FEVER

In cases of all kinds of hot fevers, give him three drops in a well distilled St. Johnswort water or Cichorii at the beginning of the Paroxismi. Early in the morning on the following day, again give him three drops in good wine on an empty stomach.

AGAINST PEST

In the case of pestilence give the patient seven drops in a good wine, and see to it that the infected person is all by himself, and caused to sweat. Then this poison will, with Divine assistance, do him no harm.

FOR THE PROLONGATION AND MAINTENANCE OF A HEALTHY LIFE.

Take and give at the beginning and entry of spring, when the sun has entered the sign of Aries, two drops; and at the beginning with God's help, be safe and protected against bad health and poisoned air, unless the incurred disease was predestined and fatally imposed upon man by the Almighty God.

But we now wish to proceed to the Oleum Antimonii and its Power, and show how this oil may also help the diseased and imperfect metallic bodies.

Take in the Name of God, very pure refined gold, as much as you want and think will suffice. Dissolve it in a rectified Wine, prepared the way one usually makes Aquam Vitae. And after the gold has become dissolved, let it digest for a month. Then put it into a Balneum, and distill off the spiritum vini very slowly and gently. Repeat this several times, as long and as often until you see that your gold remains behind *in fundo* as a sap. And such is the manner and opinion of several of the ancients on how to prepare the gold.

But I will show and teach you a much shorter, better and more useful way. Viz. that you instead of such prepared gold take one part Mercurii Solis, the preparation of which I have already taught in another place by its proper process. Draw off its airy water so that it becomes a subtle dust and calx. Then take two parts of our blessed oil, and pour the oil very slowly, drop by drop onto the dust of the Mercurii Solis, until everything has become absorbed.

Put it in a vial, well-sealed, into a heat of the first degree of the oven of secrets, and let it remain there for ten days and nights. You will then see your powder and oil quite dry, such that it has become a single piece of dust of a blackish grey colour.

After ten days give it the second degree of heat, and the grey and black colour will slowly change into a whiteness so that it becomes more or less white. And at the end of these ten days, the matter will take on a beautiful rose white. But this may be ignored. For this colour is only due to the Mercurio Solis, that has swallowed up our blessed oil, and now covers it with the innermost part of its body. But by the power of the fire, our oil will again subdue such Mercurium Solis, and throw it into its innermost. And the oil with its very bright red colour will rule over it and remain on the outside.

Therefore it is time, when twenty days have passed, that you open the window of the third degree.[17] The external white colour and force will then completely recede inwardly, and the internal red colour will, by the force of the fire, become external. Keep also this degree of fire for ten days, without increase or decrease. You will then see your powder, that was previously white, now become very red. But for the time being this redness may be ignored (is of no consequence), for it is still unfixed and volatile; and at the end of these ten days, when the thirtieth day has passed, you should open the last window of the fourth degree of fire, Let it stay in this degree for another ten days, and this very bright red powder will begin to melt. Let it stay in flux for these ten days. And when you take it out you will find on the bottom a very bright red and transparent stone, ruby colored, melted into the shape of the vial. This stone may be used for Projection, as has been taught in the tract on Vitriol. Praise God in Eternity for this His high revelation, and thank Him in Eternity. Amen.

---

[17] The alchemical ovens had small openings at different heights, by means of which the heat was regulated.

ON THE MULTIPLICATION - LAPIDIS STIBII.

The ancient sages, after they had discovered this stone and prepared it to perfect power and translation of the imperfect metals to gold, long sought to discover a way to increase the power and efficiency of this stone. And they found two ways to multiply it: One is a multiplication of its power, such that the stone may be brought much further in its power of Transmutation. And this multiplication is very subtle, the description of which may be found in the Tract on Gold.

The second multiplication is an Augmentum quantitatis of the stone with its former power, in such a way that it neither loses any of its power, nor gains any, but in such a manner that its weight increases and keeps on increasing ever more, so that a single ounce grows and increases to many ounces.

To achieve this increase or Multiplication one has to proceed in the following manner: Take in the Name of God, your stone, and grind it to a subtle powder, and add as much Mercurii Solis as was taught before. Put these together into a round vial, seal with sigillo Hermetis, and put it into the former oven exactly as taught, except that the time has to be shorter and less now. For where you previously used ten (alii thirty) days, you may now not use more than four (alii ten) days. In other respects the work is exactly the same as before.

Praise and thank God the Almighty for His high revelation, and diligently continue your prayers fir His Almighty Mercy and Divine blessings of this Work and Art as well as His granting you a good health and fortuitous welfare. And moreover, take care always to help and counsel the poor.

LAVS DEO OMNIPOTENTI

THEY MADE THE PHILOSOPHERS STONE

By

Richard Inaglese

A Golden Manuscript, with an introduction
By Frater Albertus

INTRODUCTION

Those interested in metaphysical and esoteric literature are most likely familiar with the names of Richard and Isabella Inaglese. Their books have encouraged many to delve deeper into what is called "The Mystery of Life." Among their publications still available today are: "History and Power of Mind," "The Greater Mysteries, "Astrology and Health," "The Evolution of God and Man," "Fragments of Truth," and "Occult Philosophy."

Undoubtedly those who have read the books of Richard and Isabella Ingalese suspected that the contents thereof were an outcome of prolonged study, but the direction that study was to take was almost unknown. Had it not been for a public lecture given by Richard Ingalese at the end of the 1920's (he copyrighted this essay in 1928 under his name) we would as yet be relatively unaware of the goals their studies achieved. As will be seen in this article, all their mental efforts were aimed at a substantiation of the law of polarity. It became apparent to them that theory alone without its counterpart in practical manifestation, cannot endure for long. Just as body and mind are a unit animated by spirit, so must theory become animated by praxis. Even if animation were possible, a body without a mind is incomplete and could at best be described as an automation. Yet even here, mind must necessarily control the material manifestation.

The student enters into metaphysics or esotericism by way of theory, which theory must eventually prove itself in practice. Manifestation can only be established by practical application of theory through knowledge. The outcome thereof either proves or disproves the theory. All manifestation is accomplished by the utilization of Will, which is but another term for "being alive."

Life is wherever substance exists and wherever life and substance are, there we find governing mind or consciousness. In other words, consciousness exists in the atom and electron as well as in the galaxies of the universe. Understood in this light, the old hermetic axiom "as above so below, as below so above," takes on a more profound meaning. Since time immemorial men have tried to find the key which would unlock the mystery of all existence-not just concerning mankind, but for all substance and non-substance alike. Therefore subjective and objective have separate meanings. The former dealing with the intangible and the latter with the tangible. An interplay of the two creates phenomena. Since mind cannot be separated from the very substance it governs, one cannot exist without the other.

In the universe there is but one primal substance, no matter how dense, how coarse, or how subtle and seemingly imperceptible. This primordial first substance is known as "Chaos." Substance is found in all forms and manifestations in this state and it is therefore termed "chaotic." Matter is formed out of this state according to the requirements and specific circumstances in the ever evolving spiral of evolution. In this way matter is ever becoming or ever changing and does not remain stationary. This is the meaning of the words "chaotic condition" and is not to be confused with the generally accepted definition of disorderliness. Everything in this chaotic state is according to law and order, but because of its multitudinous expressions, understanding of the true meaning escapes the finite mind of man conveying to him the image of a chaotic condition without law and order. Paradoxically, the reverse is true.

Similarly in the universe there is but one mind or supreme consciousness known also as the Oversoul or Overself, etc. Each particle of the "Chaos" is imbued with a segment of this overall consciousness, or soul essence; often called, "a spark of the Divine."

There is only one universal life or spirit. All substance is alive-even that which we call dead. A corpse for example, could not undergo putrefaction or what man calls disintegration from one state of being to another were this not the case. The overall consciousness or sum total of a coordinated being, such as man, may have its dominating composite consciousness removed and implanted in different forms of substance, but the inherent consciousness of each cell is still active as a unit. Whenever a unification of individualized particles of substance occurs, a superior state of consciousness takes temporary control. In this manner an adeptship is eventually revealed in a personification evolving through individualized cell expressions. In the animal world these cellular units of expression are known by a variety of names from the lowest amoeba to man.

Body, Spirit and Soul are the three essentials in our universe. All esoteric approaches have as their aim the separation of these into distinct entities. Whoever cannot separate Body, Spirit and Soul, is not prepared to enter into the realm of the alchemist. It is here and here only that the reunification takes place.

To test this theory was the aim of Richard and Isabella Ingalese. Only by so doing could it be substantiated or refuted. For them Alchemy provided the only means of verification.

When one reads of the time consumed to achieve demonstrable proof, it becomes evident that practical application is more difficult than theorizing. First proper reading material had to be found which would shed light on the subject and indicate a method of procedure, beginning with theoretical instruction. Even here complications arose. It was more difficult to find a competent teacher than at first anticipated as even the theoretical teachings conflicted. In fact, most of the teachers believed physical laboratory alchemy to be nonexistent. The subject matter was to be understood in a spiritual sense only. When questioned, the answers inevitably were that alchemy was of a mental nature, never to be understood literally or demonstrated on a practical-physical plane of awareness. The practical plane found few followers, while the mental had most of the adherents.

If practical alchemy lacked only proof, one may logically question why a division was necessary. This is exactly where the trouble began. There were and still are many advocates of mental or spiritual alchemy who after attempts to find a practical laboratory alchemy came to the inevitable conclusion that there was none. Why? They confused alchemy with chemistry! True, the Encyclopedia tells the reader that alchemy was the forerunner of our present day chemistry, and chemists long ago dispensed with the silly superstitions embedded in alchemy. For it is a fact that anyone taking the alchemistical symbolical terms at face value, is doomed to failure. For example, such terms as Sulphur, Salt and Mercury, are not what their names imply but are only indicative of symbolic expressions.

Here was the reason for their failure! Failure will continue for all who are not guided in factual alchemy by competent teachers. It was this instruction that the Ingalese's lacked. Nowhere is there an indication that either of them had such a teacher. The two volumes of Paracelsus translated by A. E. Waite into English, was their theoretical fountainhead. Paracelsus gave practical advice, but through the printed word only. There was no oral instruction nor the assistance of practical demonstrations. Every step had to be concentrated upon, then carefully tested, with failure as an inevitable result of their first trials. It is not surprising that many years were spent in trial ancl error before the first meager results appeared.

In 1911, Richard Ingalese then in his fifty-sixth year and his wife in her forty-eighth year were, in his words: "determined to put our conception of the teachings of Paracelsus to laboratory tests and commenced our experiments." It took them nine long years of continuous hard work before their labors finally met with success and they achieved their goal, namely the Philosopher's stone. In the first six of those years they experienced many failures and heartbreaking disappointments.

Those who have worked in practical alchemy will be somewhat startled if not disappointed by statements Richard Ingalese made in his lecture. First, he differentiated between the metallic alchemy and the medicinal, which later they decided to follow. The question is: "Why did he not test the stone upon the metals to see if it tinged"? Here would have been sufficient proof of its medicinal virtue and strength. Secondly, he stated the dose which both took twice a week, was the size of an uncooked grain of rice. This would seem an enormously large dose judging from reports of earlier Alchemists.

In answer to the first question, it would be logical to assume that the test was probably made upon metals but very wisely not mentioned in the lecture because of the consequences such a statement would have aroused in public. For this reason the question may have been deliberately avoided and left unanswered. This is only an assumption on my part, and may be taken by the reader for what it is worth.

The second question is not so obscure. The large dose could possibly have been due to an insufficient potency of the stone. Ingalese himself says that in its first state it looked like soft white marble and only after laboring an additional three years did they complete the red stone which he describes by saying, "the product was crude." This would seem to indicate that the stone lacked sufficient maturity to allow a reduction in dosage.

We are not greatly concerned here with the individuals known as Richard and Isabella Ingalese nor to their whereabouts up to the nineteen thirties and thereafter. We are, however, interested in the message they gave and in the alchemical work performed by a married couple. The husband, Richard Ingalese, attorney at law, found it necessary to give a public address on Alchemy outlining their combined results in the laboratory, and then proceeded to publish those findings.

Unfortunately, unlike Volpier, who died in 1947 in Germany, and who described in detail the process he had followed, the Ingaleses did not give specific outlines of the procedures involved. Consequently their steps and methods cannot be duplicated and verified. Even here, however, pronounced differences in the procedure are evident and cannot be overlooked. We plan to publish a translation of the original manuscript of Volpier (his nom de plume) in a future issue of the Golden Manuscript Series. Archibald Cockren, who began in practical alchemy at about the same time as the lngaleses, in the somewhat sketchy description in his book "Alchemy Rediscovered and Restored"[18] gives more details than revealed by Ingalese in his lecture.

As for now, we can only concur with the final words of Richard Ingalese in his lecture when he said, "This is our testimony on behalf of Alchemy-which each person may accept, or reject, according to his conviction."

-Frater Albertus

[18] Volume 34 of the R.A.M.S. Library of Alchemy

ALCHEMY

Wherever there is sunshine, there is shadow. Wherever there is the genuine, there is the imitation; and nowhere, in all history, is it truer than in regard to Alchemy. Particularly was it noticeable during the Middle Ages. Then there was a recrudescence of Alchemy; and because there were a few genuine Occultists that came forward to call the attention of the world to the ancient art, immediately there sprang up hundreds and hundreds of pretenders; and this has continued from the time of Paracelsus to the present. Because of these pretenders I want to talk very plainly tonight about them in order to save your faith and your pocketbooks.

By faith is meant that as soon as he has a taste of psychism, his mind turns to the Occult Sciences and then he is in a current of thought that draws to him both good and bad. I mean, by bad, those people who have studied Occult Sciences, acquired a smattering of knowledge, and, having failed to achieve any degree of success, commence to recoup all the financial outlay they have made from their credulous fellow students.

The destruction of faith is worse than the depletion of the pocketbook; but students are liable to both when they enter the occult current.

If you find that you are being imposed upon, when with all sincerity you are devoting your time, thought, and money to study, the shock is so great that you are too apt to throw the whole thing aside and say, "It is so honeycombed with dishonesty that I don't wish to have anything to do with it." That is the destruction of your faith and is a real calamity, because sometimes several incarnations must pass before you come to the point where you are once more willing to again make the venture. And so I warn you of what you may meet in the occult currents.

The Occult Sciences are the hidden sides of the physical sciences. Everything that has a manifestation in the physical world has a corresponding manifestation in the metaphysical world. To illustrate, in the case of Astronomy: If you are studying it from the physical side, you have a number of theories to account for the origin of planets. You know something of their chemical constituency, something of their movements and other incidental matters of that kind. But the occult side of Astronomy would be to know how the planets came into existence, the cause of their motion, and the purpose of their being.

The Occultist is not satisfied with theories. He wants facts. He is not content with the phenomena of life, he wants the noumena[19], or cause; and therefore he studies on both sides of all the sciences. When you commence to study causes, instead of effects, you immediately get into the occult current, and sooner or later you meet both wise and unwise people. A great many persons have studied occult books and therefore call themselves Occultists; but they are only book Occultists - quite different from practical Occultists. The sciences and philosophy are only theories with them, which may be right, or which may be wrong. They seldom attempt to prove either. These teachers of book Occultism are doing good work if they do not pretend to have what they do not possess. If they tell you they have gained knowledge from the inner side of being, and have only a theory to offer you and that theory misdirects you, then your faith is shattered.

But there are people in Occultism, as elsewhere, who teach for money only, and are indifferent to the accuracy of their teaching. But do we not find much said about such teachers in the Gospels, also?

A short time ago I received a letter from one of the most prominent astrologers in America, stating that he was compiling a book of occult formulas. He asked me to subscribe for a copy. I

---

[19] Things as they are in themselves, as distinct from things as they are knowable by the senses through phenomenal attributes.

replied, "The price you charge is remarkably low, only twenty-five dollars a copy. If your formulas are the results of the laboratory, and not of the library, you may put me down for a book; but if they are of the library, I don't want it - because probably I have all the books you have used in your compilation and perhaps a few more." He was honest enough to write, "It is the result of my researches in the library." Many persons who did not know the distinction between library and laboratory paid the price for that book, which was useless to them.

When in New York, two years ago, I met some of my old students whom I had not seen for many years. A group of them were studying Alchemy. I was somewhat amazed and, of course, interested. I asked to meet their teacher, because, for the first time in the history of the world, laboratory Alchemy was being taught openly. When they tried to find him, he had disappeared, but not until he had collected his fees. He had selected for his pupils those who had studied Occult Philosophy for years and supposedly knew something. First, he taught them what the books said about the theories of Alchemy. Then, two nights before he left - not telling them he was going - he said, "I shall give you the formula for making gold, which is easy to do; the only difficulty is to dispose of the metal after you get it." And they were credulous enough to believe it.

He said, "I will give you the name of all the elements except one. That is not permitted to be given out, but I will concentrate upon it and you who are intuitive will get that name. Then you can go home and make all the gold you want on your kitchen stove." And they accepted his statement as true.

He charged a large amount of money for his course of lectures, and the only "Alchemy" he knew was the alchemy of human nature. Of course, a person with common sense would say, "If this man knows how to make gold, why is he going about giving lectures at several hundred dollars a course, when all he had to do was to use his kitchen stove and make all he wanted?"

The person who divulged the secret is a promoter of railroads and accustomed to handle large financial transactions; and yet he was credulous enough to accept a floating faker's statement when it came to the Occult.

A woman who studied with me for a short time - a very short time - before I ended her studentship, subsequently went to Arabia; and when she returned, sought people in New York City and in Chicago who were interested in occultism - she is operating in Washington now - and said that while she was in Arabia one of the great occultists there imparted to her the knowledge of how to make the "Great Elixir."

She offered to sell it for a considerable sum, claiming that it would restore youth in a few months; and she made people believe it. Her stay in each city was limited, of course. The more incredible such statements are, the greater number of people believe them.

A man came to me a short time ago and said he had a way of making jewels. I replied, "A great many chemists can do that." He answered, "I make them alchemically. You can't tell them from nature's gems. I want you to put some money into the manufacture of them." I asked." What is your process? Just give me an intimation of one of the leading ingredients. Do you use mercury?" "No," he said. "No mercury." I could but reply, "Then you have not the knowledge you claim, for the oil of mercury is the basis of all jewels."

Other people may come to you who are earnest and sincere, but self-deceived. I was in the Calkins Chemical Company a few days ago and was talking with the manager, who said he had just had a funny experience. A man came in and showed him a mass of melted, colored glass, and said, "I want five hundred dollars, and, in the course of a month, will make myself and you wealthy men, for I have found the great art of the Alchemists. I have learned to transmute mercury into gold."

The manager happened to be a hard-headed business man, more interested in business than in the Occult, and replied, "How do you do it?" His visitor answered, "I take a pound of mercury and go out into the sunshine, get a certain angle of the sun's rays, let them pass through this glass and fall upon the mercury, and the action of those rays through this glass causes the mercury to change its vibrations; and immediately it is transmuted, before your eyes, into gold." The manager asked, "Have you two dollars?" The man replied, "Yes." "Then you don't need five hundred dollars, for I shall be just as friendly to you as you have been to me. I will sell you a pound of mercury for two dollars. Take it out on the sidewalk and turn it into gold. That will give you half your capital, buy another pound and you will have your five hundred dollars."

Twice I have been invited into the foothills near Los Angeles to see men who had Alchemical laboratories, so-called, who desired to extend their plants and only wanted a few thousand dollars for that purpose. The first man had quite an elaborate assayer's outfit and he went through the process of assaying for gold. He supposed I knew nothing about his process. After finishing, he took out some gold he previously had put in, and said, "You see how I make it?" I had to say, "I think you are a first-class assayer, but not an Alchemist."

Another man had a laboratory and wanted only ten or fifteen thousand dollars to enlarge his plant. He claimed that by passing a current of electricity through mercury it would be turned into gold. He said he had done it, and yet he was collecting money to make gold.

There are other people self-deceived, just as dangerous as those who try to deceive you. I remember a number of years ago a man came to Chicago and interested some bankers in his process of transmuting junk into gold. Some of the precious metal had to be used in the transmuting. As far as I could learn, there was no accretion of gold; but the bankers thought there was, and furnished the two hundred and fifty thousand dollars for the Alchemical Laboratory. Four days after the work was started, the inventor blew himself and his plant to bits. The man was sincere, but had a wrong theory and knew so little of chemistry, or of Alchemy, that only disaster could occur.

Forty years ago I first read the "Hermetic and Alchemical Writings of Paracelsus." Of all the books I have seen on the subject - and I have seen many - there are no others which contain so much knowledge as those two volumes. Dr. Waite's collection and translation is the best. Soon after I had finished reading the two books, a man came to me and said,

"I have been doing some chemical work, and when I cleaned my ovens I found a piece of gold the size of a silver half-dollar. I want you and your friends to join me in finding out where that gold came from. Unfortunately, I don't know. I want money enough to live on until I can find how it was made." I asked if he had no way of checking up his process. He answered, "No." I suggested the possibility that he had put something into his oven that had contained gold and when the experiment was finished the precious metal was left at the bottom. I did not join him, but many of the students of Occultism did, and lost much money.

Good strong characters who have failed in Alchemy do not defame the art. Some of the wisest men in the world studied Alchemy and failed. Robert Boyle, the great chemist, spent much of his time studying Alchemy. He was one of the founders of the Royal Society in England. He was a strong character. Finally, at the end of his life, he said he believed in Alchemy, absolutely, but that he did not have the peculiar type of mind to lead him to success.

Sir Isaac Newton spent the early part of his life trying to become an Alchemist; and when he failed, he concluded that he did not have the talent to gain the knowledge, but believed in it to the end of his life. These men are of real merit, real scientists, men of character. When the average man

318

of little stamina and petty mind fails, he turns against the science and either declares —"There is nothing in it" or, if dishonest, he goes forth to recoup his losses at the expense of the public.

From the illustrations I have given, you can see readily that the path of the investigator of the occult Sciences is beset with dangers from both wiseacres and conscious frauds. It behooves one, then, to be on guard constantly when seeking either a teacher or a companion in one's studies. Always look up the antecedents of a would-be associate. Find out what he has studied and achieved. No one reaches success in Alchemy who first did not master other things; for he requires the self-reliance which comes from many conquests before he has the "will to do" - to persist. If a person poses as a teacher, ask for some evidence of his knowledge before you enroll as his student. If he seeks financial aid to prosecute, or to complete, occult investigations, require some demonstration of his ability in that direction. No honest man could object to such requirements. A bank would not lend money to a man to enlarge his business until he had shown his qualifications to succeed. And above all, remember this, that laboratory Alchemy is never taught. It is a matter of individual conquest. It is true that after a student has shown his persistency and evidenced his character under the trying circumstances that the novice always encounters in Alchemy, and has acquired even a crude success, then some experienced ego in the great art will give, from time to time, the younger student some helpful hints which may aid him in his quest.

It requires a peculiar type of mind to succeed in this master art. I do not mean a superior mind, necessarily, but one that is tenacious, patient, intuitive, and insatiate for knowledge. All Alchemists are Occultists, but all Occultists are not Alchemists, for many students do not care for this study. Many prefer art, literature, music, sculpture, mathematics, mechanics, or some other phase of knowledge, and, in time, become masters of their selected art or science.

We speak of two kinds of Occultists, the practical and the theoretical; so there are two kinds of Alchemists, the laboratory Alchemist and the library Alchemist. The latter claim that all Alchemy is symbolical. This theory originated in 1850 when an English woman published anonymously a book entitled "A Suggestive Inquiry concerning the Hermetic Mystery and Alchemy, Being an Attempt to Recover the Ancient Experiment of Nature." Soon it was followed in 1865 by another book to the same effect, by a Mr. Hitchcock, in America, entitled "Remarks on Alchemy and the Alchemists."

These persons have many followers in both countries today. But for untold thousands of years all Alchemists, both library and laboratory, asserted that the science was a material one; and history shows that it gave birth to both chemistry and physics. The history of Alchemy also shows that at different periods there were men who acquired great fortunes without any other means of acquisition except Alchemy and who claimed that the wealth came through their knowledge of how to transmute base metals into gold. This, of course, shows that the hermetic art always has been a physical science in addition to a philosophy. Dr. Waite in his very entertaining book. "Lives of Alchemistical Philosophers," has given brief biographies of many historical Alchemists which confirm this statement.

The medieval Alchemists had to phrase all their hermetic writings in theological terms for self-protection. This, naturally, led many persons, who thought in symbols, to believe that Alchemy was mental, or, as they said, spiritual, rather than material. And such students first study and then teach Alchemy only as a means of evolutionary unfoldment. A symbolic mind can use almost any picture, or symbol, on which to hang any philosophy; and alchemy lends itself readily to symbolic interpretation by reason of its inherent nature. One of the most prominent leaders of the Theosophical Movement in Germany once called on me in Chicago to discuss metaphysics. He demonstrated his conception of Occultism by mathematics, starting with a point, then a line, and afterwards, the circle. He had read extensively and thought deeply, but took this method of demonstrating his conclusions.

It is such minds that claim all Alchemy is symbolic. These people say they are studying, or teaching, "spiritual" Alchemy. What is spiritual Alchemy? Many persons arc inclined to say a man, or woman, is spiritual when he, or she, is very thin. If persons are anemic, they are particularly spiritual; or if they adopt certain diets and reduce their weight they are spiritual, or like a spirit, or a ghost. Half of the people in the world who use the word spirit, or spiritual, have no conception of the idea behind it. It does not belong to any particular cult, or church, or to obeying the "thou shalts" and the "thou shalt nots." Nor is it subscribing to some particular creed, or adopting some theological dogma; nor is it reading certain books or participating in certain ceremonials. Spirit is the Universal Mother God, and spiritual is that which has the attributes of Spirit. What are the attributes of Spirit? Three, and three only - Omnipresence, Omniscience, and Omnipotence. An individual could not be omnipresent, because only Deity, Itself, is that; but a person may be spiritual just in proportion as he manifests in his life something of omniscience and omnipotence something of knowledge and of power. So when they speak of spiritual Alchemy there is the intimation that it gives knowledge and power. In this sense they are correct. But as generally used, that term is intended to convey the idea that Alchemy is never material, but only philosophical.

The Hermetic Philosophy, or Alchemy, started one hundred and twenty-five thousand years ago in Lemuria when the Lesser Gods revealed the knowledge to the most advanced men of that race. It did not commence in Egypt, as so many persons believe. Before Lemuria sank beneath the waters of the Pacific, it was carried by the cream of the Lemurians into India where it has been practiced, by the few, ever since. But it was revealed to the elect of the Atlanteans, also, who carried the art with them into Northern Africa just before Atlantis was submerged, and the Egyptians were the heirs of this knowledge. As India became decadent, the best of the race travelled westward and met the custodians of Atlantean knowledge, where the knowledge of the two races was combined.

When intellectual darkness settled on all the nations, Arabia was the custodian of the sacred fire that kept some knowledge and wisdom in the world. It is to Arabia, then, that almost all Alchemists gratefully must look. This should not be taken to mean that there were no Alchemists outside that country during the Dark Ages, for there have been solitary ones here and there. The word Alchemy is from the Arabic, "Al" meaning the, and "kimia" meaning infusion, or elixir; for the primary purpose of most Alchemists is not to transmute baser metals into gold, but to find the Elixir of Life. In other periods of history, other names were given to the art; but the inspiration of the great adventure has always been to control sickness and death, and all who have gained their goal have been rewarded, more or less, by this power.

Metaphysical, or philosophical, Alchemy, under whatever name it may be designated, contains certain cardinal principles, the first of which is the unity of the universe, which is one in essence. It is atomic, primarily, having two aspects, the consciousness and the material sides, the latter being the vehicle for the former. Out of this essence came, force and substance, mind and matter, in all their multitudinous manifestations. This primal essence is represented in the books as mercury; for the Alchemist learned ages and ages ago, by chemical experiments, that mercury is the mother of all things and that even life itself is but subconscious mercurial gas in motion.

The second principle of the philosophy is that there is but one purpose in the universe - to evolve minds out of consciousness, through many forms, and to develop the minds thus made into higher and still higher grades. At times this was taught openly, as evolution through reincarnation, and at other periods this truth was veiled.

The third great principle of Alchemy was, and is, that this is a universe of cause and effect. If these cardinal principles are accepted, then every claim of the Alchemist must be admitted to be at least logical. Some leading modern scientists have been driven, inch by inch, to accept enough of these basic propositions to no longer scoff at Alchemy, but to appreciate the pioneers in that field who laid the foundations for much of our present knowledge.

Empedocles, the Greek philosopher and alchemist, discovered, or rediscovered the four elements and named them. Zosimus, the Theban Alchemist, invented sulphuric acid; and I might go down the entire list, if we had the time. But it is enough to say that Geber, the Arabian Alchemist, in the eighth century, wrote a book entitled "Summit of Perfection,"[20] in which he disclosed the chemical

knowledge of the Alchemists of his time. In that
book is shown that those men calcinated, boiled,
dissolved, precipitated, sublimated, and coagulated
chemical substances. They worked then, as chemists
do now, with gold, mercury, arsenic, sulphur, salts,
and acids. Those Alchemists maintained then, as the
ancients did and the modern ones do, that all metals
are compound bodies having their origin in sulphur,
salt, and mercury, in differing proportions. This
book became the textbook in Arabia, and later, in
the colleges in Spain dominated by the Arabian
thought and culture. This book, still later, became
the textbook on chemistry for Europe and the world.
Alchemy in its esoteric form, then and later, was
conveyed to students only under signs, symbols, and
half-truths, leaving to the patient, intuitive mind
the interpretation of the symbols and the piecing
together of the half-truths into a complete science.

Most of the modern scientists, by reason of
their childish conceit, are still unwilling to admit
that the ancients actually accomplished their
undertakings; but feel that the moderns will reach
the ideals of the ancient Alchemists. The Occultist
must continue to smile at such vanity, knowing, as
he does, that time will justify, not only the
philosophy of Occultism, but of all the Occult
Sciences. This is not intended as a sneer at the

---

[20] Volume 9 in The R.A.M.S. Library of Alchemy.

accomplishments of the modern scientists, but as a caution to the intelligent student not to take too seriously the claims of the present-day scientists that they have all wisdom and success.

Nothing of basic importance has been discovered this century which does not confirm the fundamental teachings of Occultism. Take, for illustration, the theory of the electrical nature of matter and the method of its grouping. It well deserved the Nobel Prize, for it was a physical demonstration of the old Alchemistical doctrine, "As in the Macrocosm, so in the Microcosm."

Sir Ernest Rutherford bombarded nitrogen gas with alpha rays of radium and produced helium. This is transmutation of matter - dons differently by the ancient Alchemists, but, nevertheless, done. So, too, Dr. Aclolph Meithe, followed by Dr. Kurlbaum, passed electricity through mercurial vapor and changed a part of it into gold. Professor Nagaoka, of Japan, did the same thing. In the same year, Arthur Smits and A. Karsen, of Amsterdam, decomposed lead and turned part of it into gold. Is this not modern Alchemy? Why should any modern man, scientist or skeptic, presume to say the ancient Alchemists did not have the knowledge they claimed? Is there but one way - the electrical - to transmute metals? Paracelsus, in his books on Alchemy, shows seven different ways to produce the result for gold alone.

Remember what knowledge the ancient Alchemists admittedly contributed to the world, and then think how they accomplished their results with crude appliances and primitive chemical aids; and give them their share of credit, esteeming them, not as pretenders, but as men of honor and of science who were able to formulate from experiments the propositions which modern science confirms.

To be a successful student of laboratory Alchemy, one first must acquire the philosophy of the subject, and then live that philosophy until it transmutes one's nature and makes it conform to the ideals of the Occultist. This is not a very easy thing to do, for such ideals are higher than those of other cults and creeds, by reason of the very nature of the subject and the power it confers when success crowns effort.

The Philosopher's stone is the objective of most students; and when acquired and intelligently used, it confers physical immortality at will. This astounding statement is confirmed by my observation; for incredible as it may appear, I know of one Alchemist more than six hundred years old, and one whose age is more than four hundred, and another whose age is more than two hundred years; and all of these look and function as do men in the prime of life at about forty years. It can be seen from this that if a man's character is not good, if he is destructive in thought and evil in intent, he could, in time, through similar natures, organize a hierarchy of evil which, opposing the good, could delay evolution and limit its constructive harvest. And so, men of doubtful character are not permitted by Divine Law to achieve success in the higher realms of Alchemy.

If I had to define Alchemy, I would call it an exposition of nature's evolutionary processes. For illustration, let us use mercury once more, because you hear more about that than anything else in Alchemy - unless it is the making of gold. If you understand one globule of mercury, its nature, the forces that bind it together, and the chemical essences within it, it will unlock the entire Universe to you. Mercury is the key of the Universe, and that is the reason it is so dominant in all books on Alchemy. The man who breaks down a globule of mercury, to its ultimate, understands how the world is created. And when he makes the Philosopher's Stone, he becomes a real creator, for he has made a little world; and the process is identical in creating a Macrocosm or Universe.

This element is not called mercury always. It had different names in different languages. In the time of the Arabians it was frequently called arsenic, which is not the arsenic of medicine, but another name applied to mercury.

Alchemy is the mother of all sciences because
in it is contained the story of the creation of the
world, the story of the matter, the story of mind.
Were I going to picture it, I would call laboratory
Alchemy the illustration of the philosophy of
Alchemy. In other words, it is applied Metaphysics.

There are two branches of laboratory Alchemy,
the metallurgic and the medical. The metallurgic
pertains, of course, to metals. Primarily, it is the
extracting of metals from ores, then extracting the
essences from the metals and, finally, extracting
oils from the essences. Those are the three steps of
analytic, metallurgic Alchemy. After one has
succeeded in reducing and finding out the ultimate
nature of a metal, then one can reassemble it. So
that part of the science is both analytical and
synthetical. But instead of recreating the same
thing, one may break down the metal and find a
number of different elements and may reassemble some
of them to make something else. Alchemists have had
that knowledge for many cycles, and modern science
is just beginning to acquire and apply it.

To illustrate: Modern scientists can make gold,
although the United States Bulletin on the subject,
shows that it costs more to make that metal than it
is worth. For a long time chemists thought that gold
was an element but now they have accepted the
alchemical statement that gold is a compound. So,

instead of doing as the modern physicist does, putting mercury into a tube and passing an enormous amount of electricity through it to get a trace of gold, the Alchemist breaks down base metals and combines their essences to make the precious metals in commercial quantities. Modern science expects to do the same, and many of the brightest minds in all nations are devoting their lives to experiments along this line.

Professor Edwin Walter Kemmerer, of Princeton, Poland's financial savior, warns that it is time to face the probability of currency chaos caused by the discovery of synthetic gold. A few years ago, just after the Great War, newspapers throughout the world were announcing various discoveries of methods to manufacture the precious metal; and many nations feared that the Alchemists of Germany would succeed in making gold in such large quantities that they could pay their war debts with the manufactured, but depreciated metal.

The chemists of England gave a new turn recently to transmutation when they asserted that they were trying to change gold into tin and into copper, because the world's supply of gold seemed unlimited, while that of tin and of copper would be exhausted within one hundred years. This viewpoint is characteristic of England, because tin is an English product. But that nation does not take into consideration that the Andes Mountains of South America may supply all English deficiency in tin and copper and all other metals so necessary to the needs of future generations. Or, if our English chemists would concede for a moment that the Alchemists might have some knowledge, they would find in Paracelsus' books a process for transmuting iron into copper.

But nature has her own way of keeping her secrets which she reveals to those only who serve her in her own way; and so the modern Alchemists can afford to smile at the efforts of the modern scientists to transmute metals commercially, knowing that throughout the ages other bright minds have made similar efforts and failed. In fact, many of the present day chemists and physicists are in the same egos who in other lives made unsuccessful efforts in the same direction. And they will not succeed until they conquer their egotism and imitate nature as the Alchemists do.

It is not only in regard to precious metals that Alchemists flatter nature by imitation; they break down metals, extract their oils, and reassemble their atoms as semi-precious stones and as jewels and gems. There was not a crowned head in Europe which did not wear jewels made by Count St. Germain, for he was liberal with his presents to royalty, with whom he was a great favorite. The best that modern chemists have been able to do is to make small synthetic jewels. Everyone knows of the synthetic rubies and emeralds of the present day. Some are very well done and only experts can detect the false from the true. The modern chemist is less fortunate in making diamonds, producing only very small ones. The handsomest jewels in India today were never taken out of the ground; they are the products of ancient and modern Alchemists.

On this part of Alchemy, the things I have said I know, not from my own knowledge, but from hearsay - from other Alchemists and from Occult books and records. Mrs. Ingalese and I, so far, have only taken up the second branch of laboratory Alchemy - medical Alchemy.

In this connection, I have a word of explanation to offer in behalf of the nature of this lecture. For forty years Mrs. Ingalese and I have shared with the world some of our experiences and knowledge through our lectures and books. We have tried always to keep our personalities in the background, as our works show. But, the very nature of this lecture and its purpose require that for once I must break this lifetime rule, for otherwise this lecture would be useless. My purpose in giving it is to add the testimony of Mrs. Ingalese and myself to the truth of the claims of the Alchemist as far as our own experiences have gone. Those learned men have been grossly maligned during the Nineteenth Century and until the third decade of the present one. The only concession made now by the wise (?) men of the present time is that "the theories of the Alchemists were probably correct, but they never realized their dreams." And this statement is reiterated in books, lectures, and classrooms, without the least evidence to support its later portion. On the contrary, tradition and circumstantial evidence all confirm the claims of the ancients.

Our experience and our testimony are as follows: For years we thirsted for the knowledge of how to cure disease and to prolong life. We knew that a strong mind in a strong body is essential to this purpose, and therefore we studied the theories of every prominent school of medicine, and many of those not prominent. None of these held out fulfillments of our hopes. The nearest approach to our ideal was the Occult School of Medicine. For seven years we studied in this School, that being the time required to complete the course; and we were well rewarded for our efforts, though we were not taught how to prolong life indefinitely, or how to renew youth. But we were taught how to cure disease with herbal remedies and with the mind and Cosmic Forces. To save answering innumerable questions concerning this School, let me say that

Occult Medicine, like all the other Occult Sciences, is not taught in a school building situated in any particular place, but by graduates of the system who received their knowledge from an individual teacher and who transmit, in the manner in which they received them, the teaching "from mouth to ear."

No one is accepted as a pupil in this School who has not studied Occult Philosophy for at least a period of seven years, and who has not, in a great measure, lived what he has learned. The teacher, alone, is the judge of the qualifications of a pupil, and comes to him when, from an evolutionary viewpoint, he is ready to be taught-the life and the mental desire of the pupil attract the teacher.

Our study of Occultism and of Occult Medicine naturally brought us in contact with the literature of Alchemy. It was the one system that seemed to offer our hearts desire. Our other work and our situation in life were such that we could not essay the Hermetic Art at that time. And so we commenced to collect manuscripts and books on the subject and to save our money for the great adventure. This was continued for more than a decade. We learned all we could about the art, through literature and inquiry; but during those years, we deferred the attempt to try out the theories practically. But we made up our minds which branch of the subject we finally would essay.

We ascertained from the books that it was first requisite to study metallurgic Alchemy in order to know how to reduce metals for their oils. To illustrate: A globule of mercury is fluidic. The first thing a metallurgist does is to remove its metallic covering so as to "fix" its contents. Then the "fixed" portion is reduced to powder, which, in turn, is again reduced to an essence, and from that is extracted an oil. This oil is then crystallized, after which it is ready for Alchemical experiments. All this is much easier described than done, but it was necessary for us to have a definite idea of what we desired to do to accomplish our purpose, and the works of Paracelsus gave us this information. Someone has said, "You can destroy all other books on Alchemy, for their knowledge and more is contained in the Alchemical writings of Paracelsus."

In 1911 we determined to put our conception of the teachings of Paracelsus to laboratory tests and commenced our experiments. Our quest was the Philosopher's Stone, and not the transmutation of metals. We had to learn, however, the analytical side of metallurgic Alchemy, but went no further in that direction. We never have made gold, nor gems. That is a branch which is exceedingly interesting; and when we have the leisure, we shall pursue that part of the art. But we have seen and talked with those who claimed success in that branch, and, knowing their characters as we do, we have no reason to question their statements; besides which, modern science confirms the possibility, and our studies show the probability. But we are not convinced that the process is profitable. It is a question which looms large whether the time, money, and incessant labor devoted to this branch of the work would not produce larger monetary returns in some other field of endeavor. We are rather inclined to believe they would. But for the sake of knowledge, we will someday master that branch of the arts.

After we established our laboratory and commenced our experiments, it did not take us long to find that we had enlisted, not only in a difficult study, but in a very expensive one; and that our income would be overtaxed to meet the requirements. It was therefore agreed that I should return to the practice of law to supplement our resources and that Mrs. Ingalese should pursue the experiments. There have been women Alchemists in the past who have assisted their respective husbands in the work, but I believe Mrs. Ingalese was the first women to take the initiative in the art; and to her goes all the credit of the pioneer for the four long years of solitary effort and for the final discovery of how to make the stone. My part was to produce the means to carry on the work, to consult with my wife and to encourage her in the hours of disappointment and despair; and, later, first to assist in the work, and then to relieve her of the toil of bringing her results to perfection.

The essential theory of the Alchemists is that all metals have oils and these oils are the spirits, or virtues, of the metals. That was the first principle which confronted us, and, necessarily, it was true or false. The examination of the text-books on chemistry failed to disclose information on this subject: Interviews with prominent chemists brought denial of this theory; but I could not reconcile their denial with the fact that petroleum products seemed to indicate otherwise. I was told that such products were the results of animals or vegetable deposits; but learned by later investigation that this theory of science was incorrect, as is the theory that coal and its oils are derived from the vegetable world, but we were confronted with the fact that either chemistry, or Alchemy, was mistaken, and we had to determine the truth for ourselves.

As oil of gold was one of the four elements of the Philosopher's Stone - according to the books - we naturally commenced to reduce gold. But gold at two hundred and forty dollars a pound is an expensive thing to experiment with; and, after a while it dawned on us that the principle would be the same if we used copper at fifteen cents a pound.

So the experiments were transferred to the cheaper metal.

Three long, weary, heart-breaking years were devoted to the pursuit for the red oil of copper, with never a ray of light to bless the labor or to encourage hope. Nothing but dogged determination held us to our purpose. One night, a telephone message from my wife to come home at once as she had "it" which to us, of course, meant the oil. All speed limits were broken reaching home, and Mrs. Ingalese exhibited to me a brown substance that was hardening fast. She pronounced it the red oil of copper.

At the commencement of our endeavors, we had agreed that we would try never to deceive ourselves and would not hesitate to say what we honestly thought, because the easiest thing in the world is to believe what one wants to believe. Hard as it was, I had to say "That liquid is neither red, nor is it an oil, but it is greasy." She replied, "When I phoned you, it was a red oil; but it has hardened and oxidized."

So there was nothing to do but to try again; and after another experiment she produced the oil of copper. When we had that, we no longer cared what chemistry taught, nor what chemists believed. The laboratory had told us that the Alchemists were right.

I closed my office then, resigned from all my clubs, stopped lecturing and writing and all social duties and pleasures, and devoted my time to work in the laboratory with my wife. We thought that victory was close at hand, but found that it was still some years away. The fifth year gave us the oil of sulphur; but not until we had many fires and explosions and two asphyxiations. The sixth year produced the oil of mercury, the basis of all Alchemy. By this time we had sold all our securities and had two mortgages on our home, but had determined to continue with the work until we met with success, if it took this life and all subsequent ones. But we had all the oils required to make the Stone, and, thus encouraged, we tried to crystallize and fuse them.

In 1917 we succeeded in making the White Stone of the Philosophers. It looked like soft, white marble, and its effect upon the body was startling. We dared not try it on ourselves at first. But there was a third member of our family, a beautiful Angora cat of which we were very fond. We took a vote to see which of the three should test out that Stone, and the cat, neglecting to vote, was elected. It survived the first dose, and we repeated it on the two following days, with the cat becoming more frisky than usual. After that we tried it ourselves, each taking a dose at the same moment so we would excarnate together if it should prove fatal. But it proved beneficial and energized our bodies.

Shortly after that event, the wife of a prominent local physician died; and the doctor, knowing of our experiments and that the books claimed that such a Stone, if used within a reasonable time, would raise the dead, asked us to experiment on the body of his wife. Half an hour had elapsed since her death and her body was growing cold. A dose of the dissolved White Stone was put into the mouth of the corpse without perceptible result. Fifteen minutes afterward a second dose was administered and the heart commenced to pulsate weakly. Fifteen minutes later a third dose was given and soon the woman opened her eyes. In the course of a few weeks, the patient became convalescent, after which she lived seven years.

Encouraged by this success, we redoubled our efforts to make the Red Stone of the Philosophers, which is the one most mentioned in Alchemical writings. This effort was continuous from 1917 to 1920, when our quest was rewarded. True, the product was crude, but it answered to every test of a newly-made Stone; but it was so crude we were unable to retain the first dose and had to refine it by months of labor before it became suitable even for a weak medicine.

Afterwards, we commenced to take the Red Stone regularly twice a week. The dose, in size, was about as large as an uncooked grain of rice; in Troy weight, less than half a grain. The dose was very small, yet almost from the first, the results were wonderful, and, during a series of years, quite miraculous.

As I have remarked before, nothing is quite so easy as self-deception and to preclude this possibility we had a number of friends, including two physicians, to check the effect of the Stone upon our bodies. For many months, the symptoms were all subjective, such as renewed strength and greater endurance. Then the effects were quite patent to an observer; as, increased circulation of the blood, stronger heart, better color, greater number of red corpuscles, and other physical manifestations.

There were several elderly people whom we were under obligations to help in case our search proved successful, and we offered to share the results of our efforts with them; but being wisely cautious, they preferred to wait until we had tried out the Stone for a year. After that, our renewal club was formed and we all took the magic medicine. We called our group "The Renewal Club" because the books promised that the Red Stone, if persistently used for years, would renew and restore the physical body to the perfection of manhood, or womanhood. Seven years have passed, and all the members of the group - except one - are manifesting this truth in their bodies. The one member excepted was more than eighty years of age when she commenced the treatment. Her body was diseased; she did not follow the directions; and she finally died from the action of drugs, given to her by her dentist, which produced coma - kidney disease being one of her complications.

At first the action of the medicine upon our group was very slow because the Stone was weak; but as time passed, its power increased with each elevation, until in January, 1926, it was perfected for medical purposes. Mrs. Ingalese and I have not done as well as some of the other members of the group because of the condition we were in when we commenced the treatment. From 1911 to 1920, though having the knowledge and the means to keep our bodies healthful we did not use mind or any medicine in that behalf because, otherwise, we could not have known what effect the Alchemical products would have on us.

From a physiological viewpoint, those were important years in our lives, since our bodies had reached an age when strict attention and care were necessary to prevent quick deterioration. But, even under those conditions, our bodies now attest the power of the Stone, as all who have known us for the last two decades can testify.

The books, or manuscripts, claim that the Red Stone of the Philosophers will cure any illness, and that after one has taken it for five years one cannot contract any disease. We desired to test the truth of that statement and tried the Stone on many "incurables." The number of cases cured was remarkable, but we found it not infallible.

Aside from personal benefit, the one reason we entered upon the great quest was to know the truth about medical Alchemy, which I would summarize as follows:

The Alchemists who wrote on the subject usually did so within a period of a few years after obtaining the Stone. The marvelous work done by it, for themselves and others, stimulated their enthusiasm and warped their judgment. A careful observation over a greater number of years and a larger number of cases would have made them more accurate. These good men had no intention to deceive, but they spoke, or wrote, too soon.

Mrs. Ingalese and I both know that if the Stone is administered to a young, or a matured, person in normal health, it will prevent old age; that if given to a healthy but aged person, it stops further physical deterioration and starts him backward toward youth. From testimony of creditable witnesses and corroborative evidence, we believe that such cases reach perfected manhood and remain there at the will of the possessor of the Stone. So, physical immortality and perpetual youth are realities, and not dreams.

We know that the Stone restores virility in men at any age, and normal desire in both sexes. If a woman has recently passed her change of life, it restores all normal functions of the sex organs. But, if she has long passed that period, then, childbearing is out of the question

The Stone is an aid in acute diseases, but cannot be relied upon alone to cure since its action is too slow. In chronic cases, where there are no complications and fair vitality, its action is certain in any disease; where there are complications and low vitality, other aids are advisable. Of course I am assuming in the foregoing statements that the person using the Stone also exercises common sense in regard to eating, drinking, sleep and work. If one disregards all laws of hygiene and misuses mind and body, then one must take the consequences of one's thoughts and acts; for there is no vicarious atonement either in medicine or in morals. If a person desires longevity and youth, he must obey the rules in the Chapter on Immortality in "The Greater Mysteries."

It was the implicit faith in the power of the Stone to cure all disease, under every circumstance,

that caused their more conservative and wiser
brothers to use all aids to restore health, and then
utilize the Stone to perpetuate life, health, and
youth, for centuries.

I have been asked often if it were not the
mind, or faith, of the patient which produced the
marvelous results in the cases we have observed. I
answer, "NO," because some of them did not know what
they were taking, and others did not believe in its
power but took it as a "forlorn hope."

This is our testimony in behalf of Alchemy and
the Alchemists, which each person may accept, or
reject, according to his conviction, until such time
as our bodies, now sixty-five and seventy-three
years of age, respectively, by their youth and
vigor, will compel acceptance of our statements.

Finis.

The Fiery Wine Spirit

Basil Valentine

**From his "Chymischer Schrifften, Ander Teil"
Translated by Kjell Hellsøe**

A copper vessel has to be constructed that can be taken apart in the middle, as well as above and below the holes shown on the following drawing. These holes are located about, halfway up the retort.

Onto this, an alembic is mounted with a tube. Everything must be made of copper except for the recipient, which must be made of glass and be of large size. The latter is put into a wooden vessel filled with cold water. The top of the recipient must also be cooled by means of wet cloths or the liker and the cloths must be replaced by fresh ones in order to enhance the cooling effect. Also, the wooden vessel must have a cock, so that the water may be removed and fresh water poured in. When everything is ready, the prepared *Spiritus Vini* is introduced into the vessel through the lowest series of holes. When the level is just below these small holes, the SV is lit and is permitted to burn.

The air holes then force the mercury upward and the cold water causes it to liquefy, so that it descends from the alembic into the recipient. This is, then, the true preparation of the Spiritus Vini. Be sure not to neglect any of the details in the work of cooling and take care to replenish the aquam vitam so that its level does not burn too low, etc.

THE INSTRUMENT TO MAKE THE FIERY WINE-SPIRIT

(Note: This is only Kjell Hellsøe's conjecture on what Basil Valentine's device might have looked like.)

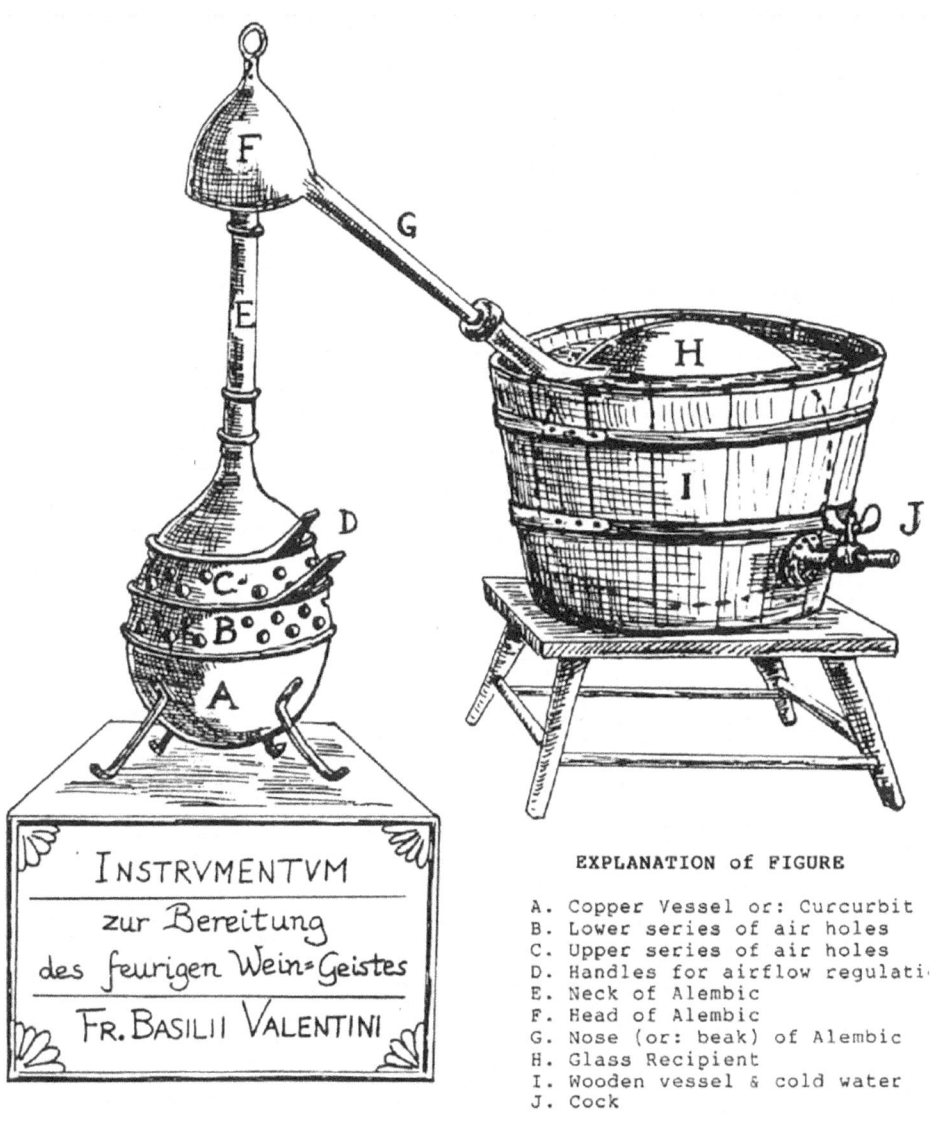

INSTRVMENTVM
zur Bereitung
des feurigen Wein=Geistes
FR. BASILII VALENTINI

EXPLANATION of FIGURE

A. Copper Vessel or: Cucurbit
B. Lower series of air holes
C. Upper series of air holes
D. Handles for airflow regulati·
E. Neck of Alembic
F. Head of Alembic
G. Nose (or: beak) of Alembic
H. Glass Recipient
I. Wooden vessel & cold water
J. Cock

Modern Apparatus to Make the Fiery Wine Spirit

(The following sketch is Kjell Hellsøe's idea of modern equipment to make this fiery wine-spirit.)

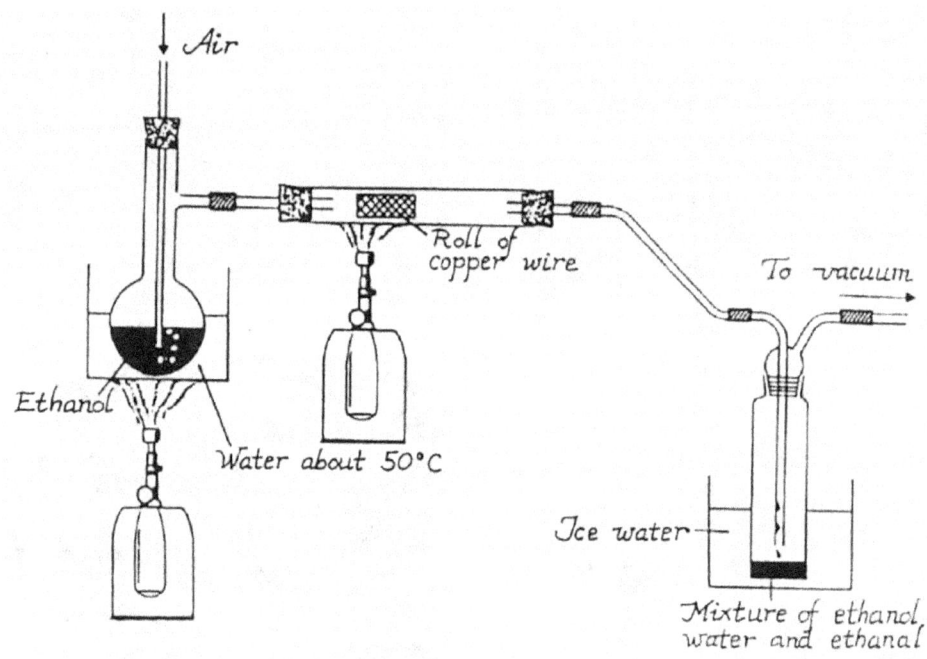

The partial oxidation of the alcohol takes place in the horizontal tube where the copper mesh acts as a catalyst for the reaction. The equation representing this is:

$$2C_2H_5OH + O_2 \quad 2CH_3CHO + 2H_2O$$

From this, it is concluded that Basilius' Fiery Wine-Spirit, is a mixture of water, alcohol and acetaldehyde, or ethanol. Acetaldehyde is a <u>very</u> volatile liquid with a boiling point, of 21°C.[21] For further information the reader can consult:

Linus and Peter Pauling, "Chemistry", Freeman, 1975 p. 424. It should be noted that the modern spagyricist can either make this Fiery Wine-Spirit, or, in fact, can purchase the acetaldehyde and mixing it with 96% ethanol. It would no doubt be of great value to experiment with this substance as a menstruum.

---

[21] 69.8° Fahrenheit.

A Word from the Publisher

Thank you for purchasing this small work from The R.A.M.S. Library of Alchemy. During his lifetime, Hans Nintzel was dedicated to the identification, acquisition, study, retyping and, when necessary, translation of what he considered to be the most important known works on Alchemy. Hans was assisted by his sparse network of fellow Alchemists, all members of the Restorers of Alchemical Manuscripts Society (R.A.M.S.). I was an active member of R.A.M.S.

My goal is to publish all of the works originally made available through R.A.M.S. as photocopies. To facilitate this, I have chosen to have the books professionally printed. I also have a few titles that I intend to add to the original R.A.M.S. Library, selected by strict criteria established by Hans.

The works from the original R.A.M.S. Library are republished by R.A.M.S. Publishing Company in the collection, "The R.A.M.S. Library of Alchemy," with permission of the Estate of Hans W. Nintzel.

If you have a work on Alchemy that you believe should be a part of the R.A.M.S. Library, please contact me through R.A.M.S. Publishing Company.

Philip N. Wheeler

www.ingramcontent.com/pod-product-compliance
Lightning Source LLC
Chambersburg PA
CBHW081431170526
45166CB00008B/2165